The Art of Technical Writing

Also by Eugene Ehrlich
How to Study Better and Get Higher Marks

THE ART OF TECHNICAL WRITING

*A Manual for Scientists, Engineers,
and Students*

by

Eugene Ehrlich

*School of General Studies
Columbia University*

and

Daniel Murphy

*Bernard M. Baruch School of Business
and Public Administration
The City University of New York*

THOMAS Y. CROWELL COMPANY

*New York
Established 1834*

To Norma and Eileen

L.C. Card No. 64-12081

Apollo Edition, 1969

ISBN 0-8152-0226-1

CONTENTS

PREFACE

This is a book about writing for scientists and engineers. It is designed to be read through when first purchased, again before the next report you write. Then it should be kept on your desk, between dictionary and thesaurus, to answer the annoying questions of form and style that crop up again and again.

A word about the book's audience. *The Art of Technical Writing* particularly disavows the attitude long held that an engineer is incapable of writing well. In that sense, this book is an optimistic one.

In the era of research, space exploration, advertising, newspapers, and paperback books, the written word is at least as important as it has ever been in the recorded history of man. It is not for us to debate the question of relative importance, but one thing can be said with certainty: much of what is being written today is the work of scientists and engineers who never for a moment think of themselves as writers. But day by day, as they push on to new knowledge and better ways of using old knowledge, they are producing an almost overwhelming torrent of words. Look at the number of scientific and technical journals now available in libraries. Think of the many times larger number of other written reports from the thousands of companies engaged in research, design, and production all over the world. No wonder journals of digests and abstracts have become such important tools.

Regardless of the undergraduate and graduate training of the scientist or engineer, once he begins his career, he is as much writer as mathematician, physicist, chemist, biologist, or engineer. As he begins to advance in his career, he becomes editor, planner of books and articles, co-ordinator of huge editorial projects. With the help of a team of associates he may turn out a multi-volume proposal for an advanced space design in a single month, and then in the next month be responsible for several new volumes on an equally ambitious topic.

No novelist of this or any other century ever wrote as much

in his lifetime as thousands of technical writers individually produce in a few years.

But the technical writer is often the man who said to himself as he plodded grimly through English 101, "Why do they ask me to sit through this miserable course? When I get my grade in this course, I will never write another page. So help me!" This book was written to provide that help.

Hopefully, those who need this book will find in its pages the direct and specific information that will make writing easier and a great deal better than it otherwise might be.

The first part of this book—Technical and Scientific Writing—deals with the types of documents that the engineer and scientist have to write. It attempts to set up usable formats and procedures for writing. It is based upon thousands of hours of teaching scientists and engineers in the Polychemicals Division of the Du Pont Company; the Linde Division of Union Carbide; the Bell Telephone Laboratories; and the Norden and Sikorsky Divisions of United Aircraft.

Chapter 6, on the management of technical writing, draws on the best practices observed in these companies and is critical of some practices that are less than satisfactory. Certainly we all have something to teach one another, and we all have a great deal to learn from one another. In one sense, the discussion of the management of writing began to be written more than ten years ago in a discussion of the art of supervision held at the great Experimental Station of Du Pont. Much of the remainder of the first part of our book grew out of long and happy participation in the Communications Development Training Program at Bell.

The second half of this book—A Handbook of Style and Usage—is a desk reference so arranged that the engineer will find solutions quickly and easily to those problems of grammar, punctuation, documentation, and the like, that so often bedevil him. The chapters were purposefully kept short so that, by using the table of contents, glossary (Chapter 18), and index, the necessary search time may be kept to a minimum.

It is customary to acknowledge the sources of assistance provided the authors of a book such as this. For the first part of our book, the list would be cruelly long, with such risk of unpardonable omission that feelings might be hurt. We wish to avoid this risk. Surely no feelings will be hurt, therefore, if we restrict our citations to the organizations already named and further mention the Education and Training Department at Bell and the Advanced Study Group at Norden. The second part of our book undoubtedly owes much to our years of teaching at Columbia University and the University of the City of New York.

We leave this book to you now, in the hope that it will prove a useful companion in the long days and nights of writing that lie ahead.

EUGENE EHRLICH
DANIEL MURPHY

TECHNICAL AND SCIENTIFIC WRITING

1

THE ABSTRACT

The abstract is the most important part of any technical report. Except for the title, it is usually the shortest part of the report. Even though it is not always specifically labeled, it is present in every report.

The abstract typifies technical writing at its best, reflecting the clear thinking and honest purpose of the scientist or engineer at work.

A good abstract discloses the most important information in a technical report. It comes to the point without fanfare, without hesitation, without the obscurity that afflicts so much writing. It puts the reader directly into the heart of the matter while setting the stage for the full and effective presentation that is to follow.

A good writer finds it quite easy to write a good abstract; a bad writer struggles but fails to write an acceptable one. The writing of an abstract surely separates the men from the boys.

For these reasons, this book, which is all about technical writing, begins with a discussion of the abstract.

DEFINITION

Some editors define two types of abstracts, the descriptive and the informative. Others call them the table-of-contents abstract and the informative abstract. This latter definition comes closer to the truth, but the best way to classify abstracts is into the *nonabstract* and the *abstract*. If an abstract does not meet its obligations fully, it is not an abstract at all.

The nonabstract is no more than a confirmation of a title. If a *technical proposal* bears the title "An Inertial Guidance System for Supersonic Aircraft," the nonabstract may begin:

> An inertial guidance system is described that is suitable for use in supersonic aircraft. Full details are disclosed

> in regard to instrumentation, research, and development problems. Technical data are presented to show the feasibility of the design. . . .

But isn't this what we expected when we read the title of the proposal? If anything but what is mentioned in this nonabstract is the bulk of the proposal, we are in for a rude shock. Why did the writer bother to write this nonabstract?

An abstract for the same proposal might begin:

> An inertial guidance system weighing less than . . . pounds and accurate to within . . . miles per hour of flight is proposed for aircraft flying at speeds up to . . . miles per hour. . . .

Now we are getting somewhere. While the nonabstract did little more than confirm the reader's expectation that the proposal he is about to read has been well titled, the abstract gave two of the most important facts that will interest us in the document. Would you rather read an abstract or a nonabstract?

The abstract tells the reader the most important facts and ideas that will be developed at greater length in the paper he is about to read. In various kinds of papers—the most important of which will be discussed fully in later chapters—the abstract meets various types of needs. Let us examine a few of them now.

TYPES OF ABSTRACTS

¶ In the *development report* the abstract tells

- what has been accomplished
- its outstanding features
- its applications, if known

Thus:

> A technique has been developed for. . . . Its advantages over existing techniques include. . . . This technique should find application in . . . and. . . .

And:

> Important improvements have been made in the manufacture of. . . . Increased . . . and . . . , as well as decreased . . . , make this development worthy of fuller testing. Existing markets should benefit from introduction of these improvements, and new markets in the . . .

industry may now be attracted by the economies inherent in the improved processes.

¶ In the *research paper* the abstract will not usually contain suggestions concerning the uses to which the research results can be put, but it should focus on

- what has been accomplished
- its most important facts and implications
- logical next steps open to study

Thus:

> Some evidence has been found that viral infections are. . . . The great majority of cases studied showed that. . . . The findings suggest that further research into the problem of . . . might consider. . . .

And:

> Vacuum deposition of certain metals has been achieved through use of. . . . Films as thin as . . . have been observed for. . . . Control of thickness remains a problem. One possible solution to this problem may be. . . .

While it is present practice to use abstracts in most formal reports of some length, many writers think that the short report, informal memorandum, or trip report is not worthy of a paragraph or sentence of abstract. Rather than tell the reader at the outset what it is that such a document is all about, the writer thinks of the reader as a captive who must read the hundreds of words the writer has put down. But a busy scientist cannot wade through every document that comes to his attention during a long, hard day. Even when he does go through a report, he may not get what the writer intended; he may even miss the main point entirely because he is tired or because the writer has written ineptly.

Whether an informal memorandum demands action on the part of the reader, whether it wants to inform him of administrative change, whether it wants to announce acquisition of a new piece of equipment—whatever the purpose of the memorandum—it should begin with an abstract.

Your company style book or laboratory practices manual may state that abstracts are to be included only in documents of a certain type, but the writer must approach any writing task with the thought that if he ever has his reader's attention, it is at the beginning of the paper he is about to write. Uppermost in his mind, then, is the question, "What do I really have to tell my reader?"

If the question is answered immediately and correctly, the resulting opening sentence will surely be a telling one. And the writer will have written an abstract even though it may not be so labeled. In the sense that the remainder of the memorandum will substantiate the opening sentence, the abstract is there.

¶ Consider these opening sentences in a *trip report:*

> The writer, accompanied by Messrs. Smith and Jones, made a two-day visit to the North Arlington Plant of the XYZ Company on January 6-7, 1963, to inspect the extrusion facilities now in operation. XYZ's plant managers, Davis and Regan, conducted our tour. We met several of their field service engineers: Hale, Crawford, Selvin, and Walter.

Come now. What did you find? Are you going to go on to a description of the plant or the cafeteria? While it is important to know the plant personnel visited, whatever did you learn? Isn't there a better way to start a trip report?

> XYZ's expanded and modernized North Arlington extrusion plant appears capable of supplying us with the polyethylene shapes we will be using in our new . . . development. Among the essential equipment now available are. . . . Shapes they are producing that are of interest to us include. . . . They promise delivery of . . . on five-day notice.

This is a more informative opening for a trip report. It answers the questions Smith and Jones and their associate were asked to study. The reader gets what he is after. He has read an abstract.

Some years ago, a neophyte chemist employed by a large company went off to hear a paper being given at a scientific meeting. His trip report began with an account of how he almost missed his train, went on to tell how uncomfortable the journey was because the railroad had routed his train over an alternative route that showed little regard for sacroiliac complexities, and finally explained that the train was so late that the chemist missed the paper he had gone to hear. While this boner is a classic case, is it any worse than starting a trip report by telling your boss that "The writer, accompanied by Messrs. Smith and Jones" actually showed up at a plant they were being paid to visit rather than see a ball game?

¶ Other *informal memoranda* to be circulated on a limited basis within a company also cry out for a few quick sentences —of abstract.

> The problem of utilizing computer facilities has long plagued the Engineering Design Group. The writer has observed that in the past month the 1410 has stood idle for protracted periods. As a result, a questionnaire was distributed.

But the co-workers who are this author's readers know just as well as he that the computer has been standing there eating up its rental fee and not turning out much work. Secondly, the questionnaire was probably instigated by the complaint of one or more of his co-workers; certainly most of them had to answer the questionnaire themselves. Finally, the subject line atop this particular memorandum read "Recommendations for Increased Use of 1410." So the reader either plods along wondering when the memorandum will get to the point or he practices his speed-reading techniques and turns the page because there is nothing on it worth reading. Perhaps there is a better way of opening the memorandum:

> Full-time employment of a programmer is recommended as a means of increasing the usefulness of the 1410. A survey just completed shows that most members of the Engineering Design Group find themselves reluctant to spend time working out routine programs that can be handled by a technician. Clerical help is also desired to save time spent hunting and pecking on a card-punch keyboard.

Out on the table, in the first paragraph, we now have the recommendations as well as the reasons for them. The remainder of the memorandum, if it bothers to go on at all, will present simply and directly a tabulation of responses to the questions asked in the questionnaire. The abstract has done its job.

¶ Not all abstracts are two or three sentences long. The reasons for making them longer are obvious.

In a full-blown *technical report* many pages long, the subjects to be discussed are highly complex. While there is an order of importance within the report, many individual topics are of great importance.

Consider, for example, a report concerning the development of a communications system involving many men over a period of many months. There are many subsystems within the over-all system, and some of them involve concepts newly developed for the system. Tests of many kinds were performed. Designs gave way to redesigns. Indeed, if experience is to be relied on in this matter, much of the system is still undergoing design changes even as the "final report" is being drafted.

The abstract for such a report must touch on all major features and innovations if the paper is going to find its greatest usefulness for all its readers. In the second half of this book, the audience for writing is considered fully. When we think of the audience for this important report, our minds run to the users of the system—and they are many kinds of engineers as well as many managers. Surely it is important to brief all these readers on the important matters that will be developed fully in the body of the report. Then there are the associates of the writer who also want to read and learn what they can from the report. They will all look for guideposts in the abstract. A one-paragraph abstract cannot include all the information these readers must have. A page of abstract is just about the right size; in 250 to 400 words a great deal of information can be given.

Some cases justify an even longer abstract.

Many companies circulate only abstracts to most of the people on distribution lists for reports. This practice reflects the finding that considerable paper and time are wasted in duplicating huge quantities of reports that will not be read fully. In such cases there is need for a long abstract, running to two full pages. It gives information beyond just what task has been accomplished, what its features are, and what applications can be suggested.

The reader confronted by an abstract circulating separately from its paper does not have the opportunity to read further in the report unless he wants to go to the trouble of ordering the full report. Again, he may not have enough knowledge of the particular technology to be able to read a complete report profitably. For his sake, then, we supply material that is ordinarily reserved for the introduction to an important report. We normally refer to such information as "background"—why the project was undertaken, what its goals and technical requirements are, what other attempts have been made to meet the same need, and the like. In this way the reader gets a capsule briefing on the subject of the report and can grasp the information conveyed in this abstract, which can properly be thought of as a digest or news announcement that is intended to keep him abreast of a late development. He still has the option of going further into the subject by ordering the full report. Naturally the full report does not include this abstract but carries the kind of abstract previously described, because it is assumed that its reader will go through as much as he wants of the full report.

More companies would do well to adopt this practice. While some circulate lists of titles of new papers published, they find that the titles are not sufficiently descriptive to satisfy many readers. Worse, many titles mislead readers into ordering papers that they eventually find disappointing. A long abstract of the kind just described would save much wasted effort, readers would order full papers with confidence, and the flow of reports directly from "in" baskets to "out" baskets without being read would diminish markedly.

¶ In a *proposal* the abstract has a formidable responsibility, so discussion of this type of abstract has been held for last.

The experts who have the job of evaluating the technical content of a proposal can probably be counted on to read faithfully through the entire proposal. They know and appreciate the amount of effort that has gone into it. But they are not the only readers to be considered. Everyone who has the responsibility for deciding who is to be awarded a contract will probably read the abstract of a proposal. This might include men who have had practically no scientific training. It surely includes men whose jobs have taken them so far from their original technical competence that they are practically laymen in the area under discussion. This kind of reader places an extraordinary burden on the writer to do his job simply and clearly in the abstract, touching on all the important points in the proposal.

But even if we think only of the expert reader, we see that a clear one-page abstract will help him find his way into the heart of the proposal.

So we expand on the opening paragraph of the abstract. The information needed in the proposal abstract is

- what is being proposed
- its outstanding features
- capability to deliver
- important related experience
- growth potential
- responsiveness of the proposal

Now we have an abstract that begins with a paragraph that is an abstract in miniature, for it touches on the first two items, the most important in the list above. But the abstract goes on to give, paragraph by paragraph, a somewhat fuller discussion of the selling points of the proposal: features, capability, experience, growth, and the scope of the response.

Nothing is said in the abstract that is not fully documented in the proposal, but nothing of outstanding importance is explained in the proposal that is not mentioned in the abstract. The reader can be thought of as saying to himself, upon completing the abstracts of all the proposals on a single subject in front of him, "If each of these proposals substantiates fully what is said in its abstract, I know now which proposal will get the award as far as technical content is concerned." Such a psychological advantage to the proposal writer who has written an effective abstract cannot be overlooked.

If we remember that in most proposals the abstract usually follows the title page immediately, how foolish it is to give up such eye-catching prominence to such phraseology as:

> The ABC Division of XYZ Corporation submits this proposal in response to Invitation to Bid No. 85-FRP-32, dated February 15, 1963, from the Bureau of Weapons.
> The program is intended to develop and fabricate two prototype ground-to-ground missile systems for use in night warfare. This development is to be carried out in accordance with specifications contained in BuWeps Document 12345, dated February 10, 1963.

If this is what the second page of a three-volume proposal is given over to, consider the effect on the reader. In the first sentence he has been told that the cover of the proposal was not in error. ABC Division of XYZ is indeed submitting a proposal; and it *is* in response to an Invitation to Bid that was issued by the reader. Surely he remembers sending the invitation. Especially since, on the day this abstract appears on the reader's desk, there also appear anywhere from one to some dozens of other abstracts in their own three-volume binders. And if he still did not realize what the proposals responded to, the cover of each proposal, or the title page, told him! The second and third sentences state that the writer really read the Invitation to Bid and knows what he is supposed to propose on. A waste of a page, compared with an abstract that begins:

> DEF Division of GHI Corporation proposes a ground-to-ground missile system that will arm the foot soldier with firepower greater than that mustered by an entire infantry company in World War II. A night-vision device and portable missile-launching equipment enable him to direct fire against targets of opportunity during night patrol. An infrared sensor is the heart of the detection and launching system.

This mythical abstract goes on to describe the technical features of the proposed design, citing weights and volume and other pertinent data, such as delivery date and capacity to produce in quantity. The page concludes with a pointing-with-pardonable-pride to similar developments by DEF Division and states that no exceptions have been taken to the Invitation to Bid.

The reader can scarcely wait to turn the page and read on. This abstract has done its job.

THE TECHNICAL PROPOSAL

This chapter will discuss the technical proposal proper. Topics normally treated in the cost section of a proposal will not be mentioned.

BREVITY

It has been said that the working drawings for an aircraft carrier would sink the carrier if they were soaked in water and then heaped high on its flight deck. Some comparable hyperbole must be available for the amount of paper consumed by United States proposal writers in a single month in the 1960's.

Anyone who has followed closely the development of a major airplane design or a project of similar scope is aware of the discouraging bulk of the paper that enters the stream of competitive bidding beginning on that fateful day when a military procurement agency announces a design competition. The bid invitation, specifications, and conditions themselves take many volumes. But they are as nothing compared with what they cause to flow from design boards, drafting rooms, art departments, and printing plants of the companies who cry tally-ho and go off in pursuit of the fox. Let us have a nightmare together.

There is the marvelous institution called the first go-around for the giant prime contractors holding their own competitions for hundreds of subcontractors. Aggressive subs bid to several primes, hoping that if they miss a spot on one team, they will find a place on another. The wily hedge their operations carefully so that they may end up on several teams at once with several different best designs. They may be on one team for one subsystem, on another for a different one. Thousands of reams and at least two postponements of the deadline later, each prime contractor has assembled his team and is ready to put the whole mess together.

What has been the writing output up to that point?

Each prime may need twenty subs on the job. (Of course, each sub has lined up outside sources for some parts of his hoped-for job, and each outside source has also written his bit.) Each sub sends along two or three fat volumes of mimeographed or multilithed material in at least ten or twelve copies each. Ten times twenty—who would care to imagine what the mail room at the prime's plant looks like on due date? Only insiders with strong stomachs can say.

And the prime has his own proposal to write. In the dream we are describing, it will easily run to a dozen volumes. When he is ready to submit his proposal incorporating the material of the winning proposals chosen from the efforts of the subs, he may have fifty to one hundred volumes to send along to the Air Force or whoever started the ball rolling. Yes, Virginia, it seems that all these volumes are in ten or twelve copies each. If there are ten potential prime contractors, anywhere from 5,000 to 12,000 volumes (10 x 50-100 x 10-12) arrive on the Big Day.

Surely this cannot be true. But those volumes have to end up somewhere. Maybe most of them are thrown away—with proper security supervision. Even so, the amount left is more than even a dream can bear.

But when the date designated for announcement of the award has come and gone with no announcement from Up There, the engineers begin to get nervous. New proposals begin to flow to the primes. The primes themselves begin writing again.

Finally the announcement comes—and not a moment too soon. The volumes have doubled again. What is the announcement? No one has won. Award of contract is awaiting word from Congress. Will the Ark ever be built?

There is some word. All but three of the primes have been eliminated from the competition. The defense people would appreciate new submittals based upon the tremendous advances in the state of the art since the scramble got under way.

Yoicks! Here we go again. Teams are alerted, new specifications are drawn up, and the writing machines crank up again.

There is no use following the dream to its conclusion. This long story is familiar. Only one point is being made right now. Wouldn't it be sensible at least to make proposals as short as they conceivably can be? At the grass-roots level of the individual writer sitting at his lonely desk, isn't it possible to avoid unnecessary repetition, to use direct language, to eliminate superfluous words?

From the time of Aristotle, all teachers of writing have urged their students to be brief. But never before the Age of the Proposal has the need for brevity been as great.

PROPOSAL FORM

The form a proposal takes is determined by the functions of the document. Taking into account all that a proposal must do, we see that a clear, well-organized document is essential. How do we proceed?

Since a proposal

- reveals a plan for work that is to be paid for by the reader himself or the group he represents
- is usually in competition with other plans for similar work
- is going to be judged on technical quality, company capability, and price
- represents the final and decisive effort by the bidder to obtain a particular contract

the proposal writer has the responsibility of making the form of the proposal reflect these requirements.

It is clear that a proposal is not a story intended to amuse, beguile, or enrapture the reader. If it were, then the best way to organize the proposal would be to begin in a way that captures the reader's attention but tells him nothing of how the story will turn out. Slowly we would build our tale until, in the final pages of the work, the reader would find the climax he is looking for—the resolution of a carefully developed story line. This process is not unlike the peeling of protective coatings of a nut until we get to the kernel we want to eat. No proposal has ever been written in just this way, not even by the rankest amateur. But many have come close to it.

Think of the reader going through your proposal. He keeps a scorecard, assigning points for good qualities and characteristics, taking points away for shortcomings. Obviously, listing all the strengths of your plan right at the beginning will help your cause. Your reader is most alert at the beginning of the proposal. His attention and interest wane as he goes further and further into your paper.

An effective organization plan for proposals has been developed to make it easy for the reader to evaluate your plan and to convince him of the technical merit inherent in it. In such an organization, the proposal is in three main parts:

- opening matter—primarily the reasons why your plan deserves consideration
- body—the plan in detail, substantiating all you have previously said about the technical content in the opening matter
- end matter—management information supporting your qualifications as bidder
  —supporting data too specialized or too lengthy to fit well in the body

Let us consider each of three main parts of the proposal in turn.

Opening matter

The opening matter of a proposal consists of

- proposal cover
- title page
- abstract
- table of contents
- list of illustrations
- introduction

¶ The *proposal cover* serves two purposes. First, it identifies the proposing organization. The appearance of the proposal cover warrants careful attention because the cover is your company's signature. Once a suitable format is selected, it should not be changed except for an important reason. All your company's proposals should be bound in the same way. In designing a cover, consideration must be given to durability, legibility, and attractiveness. A proposal may be handled by many readers in the course of an evaluation. If the proposal is successful, it will be used throughout the contract negotiations. A clear acetate sheet will protect the cover, and a plastic spiral binding will permit pages to lie flat during reading. Extravagant use of color may not only hurt legibility but also give an appearance of "slickness" that will offend some readers. In this day, however, there is no good reason for a buff student-report cover. There is a fine line to be drawn between an attractive appearance and one that smacks of extravagance, but the design of a cover must reflect company pride in the proposal effort.

The proposal cover also identifies the subject of the proposal. The title, which is printed on the cover, must be complete and explicit. *A Study Proposal* will not do, but *Proposal for Study of* . . . will. The organization to which the proposal is directed must also be named, along with information identifying the specific bid request—name, number, and date of the document. Finally, the proposal identification number assigned by the bidder must be included for easy reference if supplementary proposals, errata sheets, or revisions are subsequently submitted.

Since most of the information given on the cover of the proposal will also be printed on the title page, many companies use a cover that has a window. Thus the covers may be used for all proposals written by a company without having to overprint the cover in any way. The acetate shield will protect the window.

¶ The *title page* contents have already been described in the preceding discussion except for a statement regarding security or proprietary status of the document.

¶ The *abstract* comes next. Its contents were treated in the previous chapter, but an additional word must be said here. What has gone before in the proposal may have had a good effect or a bad one on the reader, but no decision on a proposal has ever been irrevocably affected one way or the other by the appearance or content of the cover and title page. Serious damage has been done to the reception given proposals by the content and tone of their abstracts.

As an evaluator reads your abstract, he makes his initial contact with the thinking behind the plan. If you say nothing in the abstract except what you have already stated on the title page, the reader is at least irked. If the quality of the plan is only vaguely hinted at, he will wonder why you did not come right out and say what is on your mind. If you spill the beans at once, he is able to make his first over-all analysis of your approach. "So this is what they are going to do! They make some pretty fancy claims. I wonder if they are going to be able to support them in this proposal. It is quite different from the others I have read. And darned sensible."

Trying to guess just what a reader is going to think of an abstract is as full of pitfalls as any prediction of human behavior. Some attempt must, however, be made because of the importance of the impression made by this section of a proposal. The fortunate aspect of writing an abstract is that it is short enough so that it can be completely written and rewritten many times without a great deal of trouble. Thus it can be brought to the desired quality.

In considering the content of an abstract, remember that the biggest mistakes are made by

- failing to touch on all the important points in the plan
- cluttering the page with trivial matters
- making claims boldly that turn up in the body of the proposal in such watered-down fashion that it is obvious the writer has been trying to mislead his reader
- adopting so modest a face that your strengths go unnoticed
- obscureness

The first paragraph in the abstract should state clearly just what you are proposing. The next few paragraphs will each point up one of the features of the plan. The final paragraph or two will describe the competence of the company to study, design, develop, or build, by citing past successes that have something in common with the subject of the present effort. A one-page abstract will be in line with the general policy of writing proposals that are as short as possible. One page should be quite adequate for an abstract—it is only the first of many pages of exposition.

¶ The *table of contents* must be complete, correct, and easy to read. It lists all topics and subtopics. It is both the general guide to the structure of the writing and a detailed reference to all the topics discussed.

If we think of the information retrieval problem of modern researchers, we must resolve not to add to the problem, at least insofar as the work of the proposal reader is concerned. If a numerical system is used to structure the proposal, every number used to designate a section of the report is an entry in the table of contents, along with its subject line. If a system other than the common numerical system is used, the letters or Roman numerals are given. In both systems only the page number for the beginning of an entry is given.

Correctness of the table of contents is assured by careful extraction of entries from manuscript material as soon as it is approved for inclusion in the proposal, and checking just before submission of the entire proposal for printing. This is partially a clerical problem but, like all the steps in preparation and publication, final responsibility for it lies with the proposal chief. He must see that there are no slip-ups. Because the table of contents is usually the last section of the proposal to be printed, it is especially susceptible to errors resulting from weariness and a last-minute rush.

In setting up a table of contents, the typist must take care to see that the lists can be read easily. Columnar rulings should not be used. Relationships between sections and subsections must be shown by proper spacing and indentation.

¶ The *list of illustrations* comes next in the front matter, and the same criteria that obtain for the table of contents must be observed here too—completeness, correctness, and readability.

Some proposal editors break the list of illustrations into several parts, reflecting the various types of illustrations common to technical documents. But there is no real need for such a practice. Separating block diagrams from tables, charts, photographs, and line drawings only makes for unnecessary complication. It is preferable to list all illustrations in the order in which they occur in the text. All that is needed is the number and full title of each illustration, along with the number of the page on which it appears. A word might be said here about the procedure in some companies of grouping all illustrations together at the end of a proposal. While the proposal chief saves some trouble for his printing group by this procedure, surely the great inconvenience caused the reader makes this a dubious practice. Illustrations are integral parts of the discussions that mention them, and the reader is annoyed by having to go again and again to the end of a proposal to find a figure referred to in the text.

¶ The *introduction* concludes the opening, or front, matter of the proposal. In form it is treated identically with the body of the proposal, so nothing need be said here of that matter.

The content of the introduction is worthy of close study. Bear in mind that this is the second part of the proposal to be closely read. The table of contents and the list of illustrations have been scanned, but they are not usually read closely the first time a reader goes through a proposal. He has consulted them, you might say, only to see what he is in for. He may have been sufficiently interested in one of the entries to look ahead in the paper and read the discussion of that item. Later he will use the table and list when he subjects the proposal to a second examination. He will also use them when he wants to discuss some features of the proposal with another evaluator. In most cases he will have truly read only the abstract before moving on to the introduction.

Your reader is still fresh and receptive to your argument. Writers who waste this opportunity on a dreary repetition of the material of the title page—and these same writers (see page 1) are guilty of having written a nonabstract that also repeats the title page—are wasting time and seriously challenging the reader's patience. How many times do you think he wants to be told that the proposal is in response to his invitation to bid, that your company headquarters is located wherever it is, and that the date is . . . ?

If your abstract has sketched the proposed plan and touched upon its highlights and the ability of your company to follow through on the plan, your reader is ready to follow closely as your discussion unfolds.

- Show your grasp of the central problem of the development.
- Show in broad terms how you have solved similar problems.
- Describe the company-funded research being conducted on this problem.
- Show your general plan of attack.
- Present and explain, without going into detail, the block diagram of the system being proposed.

The introduction is only a few pages long, so it will not go into great detail. The body of the proposal will do that. But after reading the introduction, the reader should have a good picture of what the bidder can and will do.

In one sense he has had an amplification of the abstract. But this is not—repeat *not*—repetition of the abstract. Each sentence in the abstract is a paragraph in the introduction, or each paragraph in the abstract is a page in the introduction. And the additional material is all informative.

If the nontechnical reader of your proposal ventures into the introduction, he may find that he does not grasp all of what you tell him, but he is a lot better off than he will be in the body of your proposal.

The technical reader now has an accurate overview of what is being offered. He is ready to go into the thorough and complete discussion awaiting him in the body of the proposal.

The body

Brevity Revisited—It is in the body of a proposal that problems of how much to write arise most often. One writer may write too much, another too little.

A scientist or engineer, proud of his work, can be pardoned for thinking that his part of a proposal is at least as important as any other, if not more important. His may not be the most elaborate part of the design, but it surely is as ingenious as any other. Besides, doesn't it take advantage of new materials, new techniques, new packaging, in a way unmatched anywhere else? Why, then, is it not natural to go on and on about it to make sure the reader will appreciate it?

Another man thinks that no one needs to be told much about what he has done for a particular proposal, that he has done more important work elsewhere, that everyone knows all about these matters already, that no one will understand them anyway. Besides, he hates to write, what he has to say has been said before, and someone else will probably cover the same material in a different part of the proposal.

The problem of co-ordinating the efforts of the writers on a proposal team is treated in Chapter 6, but it can be said here that a preliminary briefing on a new proposal is the first step in getting a large effort under way. If a team has not been briefed about how much detail is to be given in the body, what kind of information is to appear under each topic heading, who the readers are going to be, what the readers will stress in their evaluation, and what the abstract will stress, the first draft of a proposal body is apt to turn out to be a lopsided affair, too long here and too short there. Just how much detail should a writer go into in drafting a section of a proposal body?

There are several guide lines:

- Your reader *does not have* the information developed in the unpublished company-sponsored study you have been conducting.
- Your reader *does* know as much as you about the state of the art.
- Your reader *knows* that you have not actually developed the system you are proposing. If he thought you had, he would have told you so in his Invitation to Bid.
- Your reader *knows* that some of the problems that remain to be solved in the development effort are mighty tough and will take some time.

Therefore, as you write, keep in mind this reader you are writing for.

- Be brief in dealing with the routine.
- Explain in detail only the most difficult ideas.
- Let your reader know where your knowledge ends and your judgment begins.
- Don't expound on the obvious.
- Don't flog any dead horses.

Planning the Body—Think first of the purpose and scope as you plan the proposal body. What are you trying to accomplish? The reader expects you to explain in detail and to support fully the grand plan revealed in your abstract and introduction. He has the skeleton. He needs the muscle that holds the skeleton erect. He does not want any fat. In other words, he wants the whole story and he wants it quickly and crisply. What is the best way to proceed?

Your reader will understand you best if you make your illustrations the focus of your discussion and build your story about them.

There are many types of figures available to the writer to highlight, summarize, and carry much of the burden of the story. Perhaps you will want to refer to the system block diagram as a starter. You may then want block diagrams of the subsystems. Or you may want a list of system specifications. Will line drawings or airbrush renderings help? Schematics? Graphs? Tables? Put yourself in the place of the reader. What would convince *you* of the merits of the proposed design?

Put together quick sketches of the illustrations you find important. Work up dummies of charts and tables. Arrange them in the order of presentation you think most effective. Then show them to your associates for their reactions. If you are working with a proposal team, present them at the first team meeting. The comments provoked will result in some additions, some deletions, and a final plan for your order of presentation.

Beginning with the pictorial elements has two effects on the quality of your proposal. It exploits the preference of many readers for things they can see easily without close study. It helps keep down the length of the body—and this is a consummation devoutly to be wished, as some proposal evaluator once said.

Yet a picture is not worth any specific number of words,

Eastern philosophy notwithstanding. Pictures do certain things well, others not at all. The trick is to use them for what they can do and leave to words the jobs pictures cannot do.

One thing a *schematic* does quicker than words is show the flow of energy through a circuit. How many dreary circuit descriptions would be cut to the bone if they were properly oriented about a schematic! A *block diagram* not only can name system components, but it can also show their interrelations. A *drawing* can show dimensions while firmly implanting a solid impression of general appearance. A curve on a *graph* can show trends, pinpoint critical values, depict relationships between phenomena.

Why is it, then, that while engineers and scientists are fond of sketching diagrams and curves, they really do not trust their readers' ability to comprehend them? We are treated again and again to the wasteful duplication in words, sentences, and paragraphs of information that has been effectively and instantly presented in figures. Not only have the writers wasted their own time, but they have given the poor reader a needlessly bulky proposal that is apt to be read poorly and entirely unappreciated.

What a picture cannot do is discuss

- why certain components were selected over others
- why certain procedures are more effective than others
- why certain materials are better than others for certain applications
- the mathematics underlying certain proposal conclusions
- considerations important in design decisions yet to be made

In short, words are needed to convey thinking, while other symbols are available to present most data.

Organization—A proposed design will be accepted or rejected on the basis of how well it meets design criteria. The abstract and introduction have told the reader the over-all solution of the design problem. He knows broadly what you offer to do. In the body of the proposal, he expects to find evidence of just how you would go about achieving it.

The proposal body is organized in two main parts: *description* of the system and *discussion* of its characteristics. Since there is no point in reading a detailed discussion of the characteristics of a system before knowing just what the system contains, the body begins with a full description of the system. Each major component serves as a section heading and is described fully before the next component is described. Such considerations as design, construction, operation, and maintenance are spelled out, without much attention to routine details. In the description of each component the emphasis is on the most attractive aspects of the item. Indeed, not only is the best foot always put forward, but that foot is kept forward throughout the entire write-up.

After the descriptions of components are complete, separate sections will be needed for discussing matters that apply to the entire system rather than to individual components. Thus, sections are usually set up to present information such as reliability, cost considerations in design and development, life expectancy of equipment, quality assurance, ability to withstand specific environmental stress, training of maintenance personnel, and maintenance procedures. The order of presentation of these sections will depend on the type of equipment or system being proposed, but again the writer does well to consider which of these topics are the most attractive parts of the design. These sections will open the discussion.

Reams of written material are not likely in the discussions, in view of the fact that in many cases the writer can do no more than sketch the philosophy of the design as it affects these various factors. Occasionally he will be able to offer specific information rather than design predictions. Even in such cases the reader will be most favorably impressed by conciseness. When the writer wishes to cite applicable experience with similar design problems, he should do so. But such material must not be general, as it is in the introduction and abstract. Chapter and verse of previous technical solutions are called for.

Since the discussion material will often touch on details of the system design, a certain amount of repetition may creep into the proposal. If lines are clearly drawn for the various sections of the body, much repetition will be avoided. What remains can be further pared in editing.

Much of the burden of organizing the body of a proposal is often taken off the shoulders of the writer by the Invitation to Bid, which may call out specific topics to be discussed in the proposal. In some cases the Invitation to Bid also dictates the order of arrangement. It is always good practice to follow to the letter the plan set forth by the procurement agency. After all, the proposal is a document intended to sell the services of the bidder. If it is to accomplish its purpose, it must convey to the reader the feeling that the bidder is competent to do the job, has thought it through carefully, has personnel who have solved similar problems, has plant and equipment for doing the job well, and has a history of meeting development and production schedules. Certainly the proposal itself must reflect this competence by meeting the form specified in the Invitation to Bid.

Although the Invitation to Bid may specify the order of treatment of each section of the body, it will not set the form of each individual section. The bidder is left to his own devices and should give some thought to what he will do. In the interest of serving his cause best, the writer presents first those aspects of the discussion he thinks would help him most. In general, of course, this means beginning with the most important and working down to the least important. Certainly the writer does not fail to give the strengths of his design greatest prominence, both by dealing with them first and by treating them at greater length.

The best way to emphasize the strong points in a particular discussion is to begin by giving a short statement of the advantages of a particular component, for example, before actually describing how it works. The same technique holds true for other kinds of material in the body. The advantages of an analytical technique, a packaging philosophy, a particular material, or a circuit can be pointed out the same way.

In this manner the reader first finds out *why* before he finds out *how* or *what*. He is predisposed to understand and appreciate the significance of each part of the design.

However you plan each section of the body, remember that your reader does not want an encyclopedic account of every last nut and bolt. He is thoroughly familiar with the routine aspects of your subject. Use your allotted space for telling what he does not know so well.

Stylistic Considerations—In writing the body of a proposal, it is important to establish and sustain the illusion that the writing is the work of one author. The one exception to this rule comes when a section of a proposal body is actually the work of a company other than the bidder, in which case that section should merely be bound into the proposal with some identification of the company that produced it. In the remainder of the body, however, even though the writing may have been the work of a dozen or more people, the presentation represents the thinking of an organization and should reflect a unified approach to the problem. To save a lot of time, matters of style for the proposal body should be specified before any writing gets under way. Such devices as how the reader is to be addressed and verb forms to be used must be decided. Small matters, but important ones.

The most vigorous writing in a proposal body comes when the writer works under the assumption that his proposed system really exists. See how this effect is felt in these following two sentences. Which do you prefer?

> The optical system (Fig. 4) consists of a concave spherical reflector, aperture, lens, and screen, all symmetrical about an optical axis.

> The proposed optical system (Fig. 4) would consist of a concave spherical reflector, aperture, lens, and screen, all of which would be symmetrical about an optical axis.

The word *consist* could well have been replaced by some form of *to have* in both sentences, but the first sentence is surely superior to the second. The superfluous words required by the shift away from the indicative mood are reason enough for condemning the second sentence, but the shift is guilty of a greater sin. It creates a conditional attitude of *if* and *maybe* that does not help the proposal. Multiply the number of unnecessary words by the number of sentences in the body of a proposal, and you have quite a few extra pages of typing and quite a few extra pages of reading. Go on for page after page with *would* and *could* and you cannot help making the reader wonder whether your proposed system really *can* and *will*.

Of course the best way to set up the sentence in question is to have it refer to Fig. 4 properly. The sentence would then largely disappear. Surely the reader has seen Fig. 4 and believes his eyes. The first sentence he wants to read concerning the material shown in Fig. 4 is some comment on the effectiveness of the optical system depicted, its low cost, simplicity, reliability, or the like. Keep this in mind next time you write.

Chapter 11 discusses the question of verb forms more fully.

End matter

The statement was made earlier that no proposal ever succeeded or failed because of the appearance or content of the proposal cover or title page. The proposal is one book that is not judged by its cover. But many proposal writers appear to think the technical content of the book is the *only* thing that counts.

Too frequently writers concentrate their efforts on technical content and treat end matter as though it were a stepchild. So we characteristically witness writing and rewriting, splendid art work, and careful documentation in the body of a proposal, while slipshod, error-ridden, vague, dull, and uninformative pages are permitted to stand in the final part of a proposal. This is a mistake—particularly since, once written, end matter often is used over and over again as boilerplate, thus perpetuating initial carelessness.

The end matter of a proposal includes so-called *management information* and one or more *appendices*. It is not the appendices that suffer from lack of attention by writers. In fact, most writers are willing to lavish attention on these detailed substantiations of technical points made in the body of a proposal. If it is true that there is at least a little bit of the teacher in every scientist and engineer, the didactic appendix is proof enough. Therefore little need be said about the appendix here.

Mathematical proofs, intricate discussions, and the like, when considered too detailed for the reader to bother with in an earlier discussion section, are properly saved for the very end of the paper. The reader is told during his reading of the body that additional information is available in an appendix. He knows that he can find it if he wishes. It is possible that only an exceptionally diligent reader will actually consult an appendix, but the material must be there nevertheless. Omitting it would damage the proposal, but placing it outside the main stream of the discussion can only help the proposal.

It is the management sections of the end matter that suffer from neglect. The proposal writer is, after all, a scientist or engineer. He plays out his traditional rôle and cringes from anything called management. He may feel that it is not his function to interpret the management aspects of his company. Yet, while he may not be the best person for writing this part of the end matter, he must take responsibility for seeing that the writing is done as well as possible.

Management Sections—Almost every Invitation to Bid asks the bidder to supply details concerning

- company history
- company management
- company experience
- company personnel

No matter how ingenious the proposed technical solution, if a bidder cannot show that he is capable of fulfilling all the promises made and all the programs outlined, his proposal is worth little. It is in the management sections of the end matter that the bidder gives the procurement agency the opportunity to judge the ability of a bidder to deliver the goods.

Let us look at each of these management sections in turn.

¶ The evaluator of a technical proposal is not particularly eager to read a saga each time he goes through a *company history*. He cares little that Amadeo Alpha founded the great Double A complex in an unguarded moment behind the woodshed in his father's backyard. He does not want to be romanced. He wants only to be made somewhat familiar with the vital facts of your company's history. Was your company formed as a result of a decision by two Nobel Prize winners to try their hands as entrepreneurs? When was your company founded? How did its various branches come into being? Where are your plants? What facilities do you own or lease? Does your company carry on programs of research and development other than those paid for by customers? Is the company called on regularly to perform certain kinds of tasks that it alone can perform so well? Does your company have a financial statement? These are the kinds of information an evaluator wants to read.

Do you see how such facts can give your reader some appreciation of your competence? Whenever you wonder whether a certain piece of information belongs in your company history, base your decision on whether it would enhance your capability in the mind of the reader. But

- Be brief.
- Use numbers when possible.
- Use words rather than phrases.
- Use lists rather than paragraphs.

¶ Let tables of organization carry most of the burden of describing your company management. They tell the management story quickly and unequivocally. For technical departments, supply names as well as titles. The reader is impressed by the air of stability imparted by an organization chart that dares list names. Be sure to include all departments that will play a part in the proposed development.

The chart of the project organization that will be responsible for the work described in the proposal must name all key personnel. The reader must be made to feel that you have set up a team that will carry the project through successfully. Later, in the company personnel section, you will give detailed information concerning the scientists and engineers mentioned. Here the focus is on what part each one will play and how his function will be related to the others.

In addition to the charts you will supply for this phase of the presentation, you will have to write a description of how the project team will function.

But be brief.

¶ In meetings of a proposal team, the discussion inevitably turns to consideration of what *company experience* is appropriate for citing in the proposal. This matter must be decided before the introduction and abstract can be put in final form, but the end matter cannot be completed without citing experience, even in first draft.

Some engineers are addicted to mentioning every job their company has ever worked on; the lists of pertinent experience become longer and longer. Other engineers seem to feel that nothing in the past of a company is quite applicable to any new job being bid. Somewhere between these extremes lies the appropriate amount of detail.

The criterion, of course, is pertinence. What specific engineering skills are required for the proposed program? Every past job that used these skills is germane. But the term *engineering skill* is subject to outrageously broad interpretation. Use of a slide rule is not acceptable as a criterion, nor is use of a computer. What is? Does the proposal concern computer control of a manufacturing process? Have you done that sort of job before? That experience is worth mentioning. Does the proposal concern a two-dimensional tracking system? Pattern recognition? Visual displays? What have you done in these fields?

A second type of company experience is also relevant. Even though a bidder may not have a great deal of specific experience to cite on the broad topic of the proposal, he may be bidding on a program that will require a systems approach at which he excels. He may have a fine record of systems achievements to cite.

Finally, the experience of the subcontractors on your team can be cited. Because the subcontractor is an integral part of the team, he must be qualified in the eyes of the reader.

Whatever the content of the experience section, the emphasis must be upon ease of reading. This will mean lists wherever applicable and concise explanations wherever lists will not do. Of overriding importance is the need for specific information. Name names of previous and current customers. Tell what you are doing for them if it is possible to do so. Titles of contracts performed and systems previously built belong in this section of the end matter, which is organized around topics rather than chronology. Your most recent jobs are perhaps your most important, but sometimes there are older jobs that are even more relevant to the job at hand. Think of what your reader will be most impressed by, and begin with that.

¶ Statements are inevitably called for concerning *company personnel*—the size and training of your work force, recruitment procedures, and employment policies. What is needed here is the briefest possible statement that is responsive and accurate. There is no problem in the writing of this part of the company personnel section, unless the output runs to more than a page. If it does, a blue pencil will correct that mistake.

Biographies of professional personnel are also called for in this section, and they frequently become a writing problem. If all the engineers involved in a proposed program were the holders of several academic degrees, many patents, years of experience, were developers of complex and famous systems, and the like, there would be no problem in writing biographies. Unfortunately, most engineers who work on a project team do not meet these specifications. Writing their biographies is a problem.

Because engineers may come and go with some frequency, the bulk of an engineer's biography is often given over to a listing of all the companies that have employed him. Is that kind of information going to impress your reader? Will he have a picture of a team of men working together to see a project through?

It is better practice to concentrate on the engineering experience of a man rather than on the companies for which he has worked. This means avoiding the approach to biography that tells that Engineer X started his career with the ABC Company, went on to work for DEF, GHI, JKL, and MNO before coming to PQR last year. Rather, the orientation might be that Engineer X has worked on development of an A, B, C, D, and E during the past fourteen years. That he has served as project engineer on the first two and consulting engineer on the others. That his specialty is . . . and that at present he is conducting a study of . . . for the . . . program.

Of course it is important to list the academic qualifications a man possesses, including the names of the degree-granting institutions. Any patents, articles, or honors should certainly be included. But care must be taken lest some of the biographies run to more than a page while others are only three lines long on an otherwise empty page. It is far better to keep all biographies down to no more than one page and print three biographies on a page rather than make some of the team members look ridiculous.

The best form for a biography is a list. There is something deadly and slightly ludicrous in forcing a reader to read fifteen times over that "Mr. X joined our firm in . . . and is in charge of . . ." There is nothing wrong with that construction except its needless repetition. A list will eliminate excess verbiage. A few section headings can be developed that will be appropriate for most biographies:

- <u>Fields of Interest</u>—work areas to which the man has devoted himself or for which he has special training
- <u>Recent Experience</u>—specific projects he has worked on recently and the roles he has played in them; the definition of *recent* is open to interpretation
- <u>Work Experience</u>—companies and positions
- <u>Patents Held</u>
- <u>Academic Honors and Professional Affiliations</u>
- <u>Academic Preparation</u>

Notice that the order in which this information is listed reflects the appropriateness of the man's assignment to the proposed project. It does not take him along from the day he left college to today.

Boilerplate—It is common practice to prepare certain parts of the management information in such form that it can be used for any proposal a company may put out. This is excellent practice, provided the material is kept current. But some flexibility is needed if this boilerplate is to be appropriate for every proposal. Therefore, the boilerplate must be written in several forms, based on the specific needs of the types of proposals a company makes.

Reviewing the four kinds of information required in the management portion of the end matter, we find that the *company history* can stand in one form for practically all proposals. It is so rare that a company will deviate completely from its pattern of bidding that it would be better on the rare occasion to write a special company history, if necessary, to show the company history in a light different from that normally presented. This is also true of the *company management* portion. Only when a change is made in organization or when key personnel are shuffled is there need for modification.

The *company experience* section is a horse of a different color. A given company may interest itself in many different technical areas. Yet, in a given proposal, the effort is always made to show that the company more or less specializes in the kind of work being bid. This would mean, then, that several different statements must be available for inclusion in proposals. Each one will be written from a different point of view, stressing the particular part of the company's experience that will show it off best. Inapplicable experience is omitted; applicable experience is emphasized to different degrees.

Biographical information may be maintained on boilerplate for the *company personnel* portion of the end matter, as may the brief statement of personnel practices. But this information must be kept current.

JUDGING PROPOSAL WRITING

The best proposals are those that win the contract. Yet so many factors involved in awarding contracts have nothing to do with the proposal as a piece of writing that to judge a company's proposals on this basis alone is foolish.

Judging any writing is at best a difficult task. We tend to like those efforts we like, and that is all there is to it. The Educational Testing Service in Princeton, New Jersey, once set out to find out how experts judge writing and whether they agree on what good writing is. The results of the study showed that there is much disagreement even among experts.

This does not mean that a company should abandon its efforts to write better and better proposals. Whether company officials agree on what is good writing does not seem to matter. What does matter is that these officials show interest. Wherever interest is shown in quality of proposal writing, quality seems to improve.

LETTERS AND SHORT REPORTS

The engineer or scientist faced by the need for writing a technical letter, trip report, informal memo, or conference notes has as much responsibility for clarity and conciseness as he has in a more ambitious writing task. Yet, mistakenly, he often gives too little thought to the quality of these lesser documents.

¶ The *technical letter* is often meant to sell a product or service. Or it may explain some detail of procedure or function, answer a service complaint, or request or supply technical information.

¶ The *trip report* may supply information concerning customer needs or capabilities. Or it may disclose important scientific or technical information.

¶ The *informal memo* covers a lot of ground: observations on a new process or material; suggestions on administrative procedures; evaluations of potential or existing markets; requests for personnel or equipment; changes in laboratory operations —the list is endless.

¶ *Conference notes* supply a record of important meetings involving company personnel.

If any of these documents are worth writing at all, they are worth writing well.

In every organization in operation for a number of years, stereotyped forms and writing styles exist for various reports. There is nothing intrinsically wrong with having a form or company style that readers find useful. Instead of having to think, "How shall I start this thing?" the writer hauls out the printed company form and begins filling in blanks. When a printed form does not exist, he digs out the last report written by him or his boss and follows the pattern. Why should a scientist or engineer have to worry about originality of writing?

There is something wrong, however, with a form or style that is really unsuitable. If it is clumsy, if it contributes to obscureness, if it jars the reader, if it calls for unnecessary information, if it is monotonous, or if it is redundant, the form must either be discarded or radically revised.

In one large organization, for example, almost every memo begins with:

The purpose of this memorandum is to. . . .

The rest of that first sentence is a verb—report, state, explain, request, inform, and the like—followed by the title of the memo *verbatim*. Surely there are better ways to begin in most cases. But how did such a form get started? How did it ever become so popular?

Well, if you don't know what to write first, how to break the hypnotic spell of that clean white sheet of paper, you can at least make a start by stating the purpose of the memo. After all, your earlier memos did it that way, and your boss approved them. He had to, didn't he, because all his reports begin the same way. Somewhere in the dark beginnings of that old company, an engineer sat staring blankly in the same despairing way as have thousands since him. After several false starts, perhaps, he hit upon the magic formula. Others saw it, liked it, and took it up. Why fight success?

So it is that many company forms become established.

In one organization, the headings in the memo were established by a consultant on writing who was hired in the early days of the company. While the forms he set up may have had much merit when they were adopted, they are no longer understood by writers today. Semantic changes in twenty-odd years are responsible. How do you differentiate, for example, among all these items spelled out in that company's report format:

- title
- subject
- statement of the problem
- foreword
- preface
- introduction
- findings
- summary
- conclusions
- inferences

And this listing is not complete. When the originator of the form explained just what he wanted under each heading, there must have been some sound reason for each entry. Many years later there is not.

In another company, style is set by a venerable grammarian whose ideas predominate because no one dares challenge him. He believes that every statement worth its salt must be carefully encumbered with qualification heaped upon qualification, that there is no substitute for the semi-colon, and that a sentence cannot say anything worthwhile unless it runs to at least fifty words. He is editor-in-chief of every written document, no matter how brief, and the engineer who tries to change the established order is not appreciated. His report is returned to him, and he must work it over again to make it conform. A long lecture usually accompanies the rejected report. Not many will want to spend time bucking the system.

In another company a course was set up for writers to teach modern style based on a corruption of the teachings of Rudolf Flesch and Robert Gunning. Every memo written from then on had to reflect considerations of *plain talk, fog count,* and *readability level.* Intent and meaning were subordinated to the hypothetical needs of a mythical reader.

The important function of relaying information and ideas in a useful manner is harmed by all blind adherence to a "company way." Every laboratory must be constantly alert to doctrinaire beliefs that what it practices is best.

THE TECHNICAL LETTER

In estimating the costs of maintaining a modern laboratory or plant, the most important item taken into account usually is time. Capital investments in plant and equipment are major items, of course, and businesses are always seeking ways to use their facilities economically. Efficient use of engineering and research time, however, is a highly elusive art and the constant concern of managers.

This may seem a strange introduction to the problem of writing technical letters, but it is a necessary one. Engineers and scientists write many technical letters, and much of what they write wastes time. The fact is that these letters cost a great deal to write and a great deal to read.

The present discussion is not going to offer a magic formula for helping writers save a few minutes on a letter. It takes no less time to write a good letter than a bad one. Nevertheless, in striving for efficient use of engineers' time, the technical letter can play an effective role.

The trick is not to write when you don't have to.

When an inquiry comes to the attention of an engineer, he first has the job of deciding whether he can best respond by writing a letter, making a telephone call, sending a TWX, paying a personal visit, or using some combination of these methods.

- Is the request so stated that, without further clarification of any kind, a letter can supply a complete answer?
- Does the request require considerable clarification before an answer can be undertaken? Obviously, a telephone call will get that needed clarification most efficiently. And the answer can often be supplied in the same call!
- Is there need for immediate and brief follow-up, perhaps after submission of a proposal or an exchange of letters? TWX may be the best instrument.
- Is extended discussion needed? inspection of some material? tour of a laboratory? formal presentation before a group? A personal visit must be the answer.

Thus, the best way to save letter-writing time is to think carefully about whether sending off a letter is the best way to handle a problem. You will find that letters frequently come off badly among the choices available to you.

When a letter is the best solution, full speed ahead!

- Make your letter brief and to the point.
- Open with an abstract that includes identification of the letter being answered.
- Answer in order all the questions asked.
- Speak to the reader in language he understands.
- Make the letter personal and warm. You are writing to a member of the human race, not to a digital computer.
- Avoid old-fashioned and meaningless expressions so often found cluttering up business letters.
- Enclose helpful printed matter in the same envelope in which your letter is mailed.
- Close with a clear statement of the next step in the correspondence. This may consist of an offer to send additional help, an offer to call, an invitation to visit, an offer to ship sample materials, or the like.

The opening paragraph of the letter—the abstract—is the

most difficult for writers because the beginning of anything is difficult:

> Several solutions are possible for the viscosity problem you described in your letter of December 5. In order of our preference, based on what you have told us of your situation, they are:
> 1. . . .
> 2. . . .
> 3. . . .
> 4. . . .

What follows—the body of the letter—is a series of paragraphs, each describing in broad outline one of the solutions. The final paragraph offers the writer's services if more help is needed.

> If you feel that you would like to have us take a closer look at the problem and make a final recommendation, please feel free to call me. We are always ready to help.

Some writers may feel that such a letter is too abrupt. Others, that it seems too personal. They would rather it began with a long clumsy statement expressing delight over the fact that the reader referred the problem to the writer. Or they may prefer the opening paragraph to contain a statement acknowledging receipt of the original letter and stating quite clearly that the writer has read it! This kind of nonsense has been permitted to go on for much too long.

If ever a writer has the opportunity and the responsibility for chipping away at this ridiculous tradition, the scientist or engineer is that writer. You have a job to do. Get on with it.

Concern yourself with problems of clarity rather than empty phrases. Remember that your reader may not have your technical competence or degree of familiarity with your specialty. If he did, he would not be asking for help. Remember also that in speaking with your associates you have developed certain shortcuts in speech and writing that outsiders cannot grasp. Spend your writing time making clear the technical content. Your reader wants information, not empty formality.

Throughout this discussion, attention has been given only to letters that respond to inquiries. But the engineer may himself have to write a letter of inquiry. Or he may have to initiate correspondence designed to sell a product or service. Think of your experience in reading inquiries addressed to you. You will save your reader a great deal of trouble by avoiding mistakes in letters you have read.

Your most formidable job is to state exactly what you need. This is not always easy to do. Try a first draft out on your secretary or one of your associates. They will tell you whether it says what you mean. You must also tell your reader why you want the information. After all, he will not be especially eager to answer what may appear to be an idle request. An opening for such a letter might be:

> As you may know, XYZ has for some time been interested in the problem of large-scale production of. . . . In our studies, we have found that one important obstacle is selection of a suitable epoxy resin. I turn to you because of your reputation as a supplier of these resins.

At this point the reader knows generally what the letter is after. Now it must become more detailed. What qualities are needed in the resin? What special problems are there in manufacture of your product that would affect a choice of resin? What resins have you tried? How did they fail? Once you spell out this kind of information, you have only to close your letter with a brief expression of thanks for the effort you are putting your reader to. It is unstated but obvious that he will be kept in mind on that happy day when the company is ready to order a few thousand drums of resin.

Finally we must consider a technical sales letter in which the writer, uninvited, is sending information that may fill a potential customer's need. A letter that attempts to do this should be handled only by the most skillful writer, because readers are wary. They have wasted too much time reading letters that begin:

> This letter will take only three minutes of your time but it may save you. . . .
>
> Reading this letter may be the most important thing you do today. . . .

Scientists and engineers get this kind of junk mail all the time. They have become hard to sell.

What these readers want is a *data sheet,* attractive but not overpowering, informative but not so detailed that it requires an hour to read. A cover letter accompanies this sheet. It contains only two paragraphs. One tells why the sheet was sent to the reader, the other how the reader can get in touch with the writer for further discussion.

The emphasis throughout this discussion of the technical letter has been on the need for sensible organization, brevity, and avoidance of meaningless words and sentences. By becoming critics of the letters they receive, writers can improve their own letters. Look closely at the next letter you receive. How much of it can be cut? Look closely at the next letter you write. How much of it can be cut?

THE TRIP REPORT

Frequent travel is becoming more and more the lot of scientists and engineers employed in industry. Particularly is this true in companies primarily dependent on defense contracts. Trips are made in order to find out the needs of the military, to attend briefings, to maintain liaison while a project is in the works, to follow up a proposal, and to confer with other companies on a proposal or project team. Even in other fields of research and development, many kinds of trips must be made: liaison with customers, consultations, technical and scientific meetings, or plant visits. If maximum benefit is to be gained from such trips, a report must be written and distributed upon return of the engineer.

And these trip reports must be

- written promptly
- written well
- distributed promptly

In news writing a reporter answers six famous questions: who? what? when? where? why? and how? In reporting a trip, a scientist or engineer must answer similar questions. The routine questions are best handled in the format adopted for trip reports by his company. Thus, a typical trip report might be set up in this way:

TRIP REPORT

April 15, 1963

Bureau of Navigational Aids

DISTRIBUTION: all department heads
WRITTEN BY: P. D. Quirk
SUBJECT: Star Gazer application
TRIP PERSONNEL: J. L. Anderson and P. D. Quirk

Such a format makes it possible for the writer to start the body of his report with the information he has obtained, the result of his visit, or whatever is most important—for this is what his readers want to find out.

> FINDINGS   R. D. MacIntyre, chief of Bureau of Navigational Aids, has shown great interest in our modified Star Gazer and would be willing to read a specific proposal. The application he has in mind is. . . . Bureau staff engineer R. D. Morton said that initial requirements call for three prototypes for ground testing by 9/1/63, testing to be completed by 11/1/63. Production quantities will exceed 700 if their . . . program is completely funded, and the chances are excellent that it will be. Money can be found for the development in Bureau's present budget.
>
> The most important features of the design for their needs are
> 1. . . .
> 2. . . .
> 3. . . .

The writer then suggests a plan of action:

> RECOMMENDATIONS   A proposal team should be assembled without delay to study these requirements and come up with a suitable proposal within thirty days, if we are going to be able to meet Bureau's schedule. Estimated manpower requirements are. . . . Proposal chief would be. . . . Help would be needed from three groups: . . .
>
> In the initial phase, we will have to concentrate on improving . . . and. . . .

The report concludes with a section titled DISCUSSION. It describes the Bureau's program, discusses specific points more fully, and the like. It backs up the plan recommended by the writer.

When the reader has finished, he knows the most important information uncovered and the plan for suggested action. If he must help decide whether or not to go ahead with the plan, he has enough information with which to do so.

In other types of trips the purpose may be entirely different, and the report will reflect that difference. A trip to a scientific meeting obviously does not often produce the kind of information given in the sample report. But if a trip is at all important, the writer can come up with a satisfactory report that gives the stay-at-homes the gist of the information gained and its usefulness for his company. If the trip was made to investigate a customer complaint, the same opportunity exists: what was the complaint? how did the writer take care of it? is the customer happy? will some work have to be done to correct a problem? what work is suggested? should the company modify its present design or manufacturing procedure in any way?

In trip reports, then, as in all technical reports, getting to the point quickly is the writer's main consideration. Anything irrelevant is omitted. (Do you recall the case of the new chemist in Chapter 1?) Anything important is emphasized by giving it a prominent position in the report and by giving it all the space it needs. Minor points are made but clearly subordinated.

When a writer feels that he needs a great deal of time for preparing a trip report, he is probably going about his work badly. If a trip report is to do its job, it must appear quickly. Of course, in the case of the trip report to the hypothetical Bureau, the writer probably gave an oral report to some of his associates upon his return to the laboratory, but the written report is needed as soon as possible by all those concerned with the project.

The trip report can be made quickly and well if the writer gets to work on it as soon as he can. He will be able to nail down his material if he has the energy to think it through on his trip home. An outline can be prepared of the main points to be covered in the report. The writer should think of the three headings suggested for his report: FINDINGS, RECOMMENDATIONS, DISCUSSION. He may find it useful to phrase those headings as questions to himself: What have you found? What do you recommend? What support do you have for what you just said?

Writing the answers to direct questions such as these usually results in readable reports. If the writer does not have information for all these headings, he certainly must not follow the format suggested. The quality of trip reports is not enhanced by padding, and the salaries paid engineers do not often depend on the number of words used.

THE INFORMAL MEMO

Some companies—usually the very large ones—emphasize the necessity for putting facts down on paper. If a decision is needed, even on a small matter; if many people have to be in-

formed; if a permanent record is needed; if company personnel are dispersed at two or more locations; if chains of com-
mand are important—under all such circumstances a written
document is considered essential. Other companies are less
formal. They rely primarily on word of mouth, but they certainly have their share of paperwork anyway. The trick is
not to let the paperwork get out of hand. In one division of a
large company annual charges for reports of all kinds—including engineers' time and clerical machinery—come to more
than $1 million. Fewer than one hundred engineers are em-
ployed in that division! We are not discussing a trivial matter.

Much of the written matter turned out in any company
can be classified under the heading of informal memos. Because these memos include an almost infinite number of types,
they cannot all be discussed individually. The best way to consider them is by examining the requirements they have in common and suggesting a few patterns to follow.

Let us think for a moment of some of the subjects of such
memos: ideas still in the infant stages of thinking; minor technical problems in a development; a bit of information collected from a customer or a salesman in the field; a report of
trouble from a technician; the need for some new equipment.
As we think about any one of these topics, it is clear that the
memo presenting it must be brief and should be written easily.
The writer should almost be able to talk it to a stenographer,
pencil in a correction or two on the typed copy, and send the
memo off.

Why not? Such a memo does not reach a broad readership.
The writer probably has personal contact with his readers
every day. The writer knows the language they speak and the
technical background they possess. Above all, the memo is
not of earthshaking consequence.

The structure of the informal memo is simple. Do you have
an idea you want to suggest?

> The catalyst needed for the . . . process must have
> the following characteristics: (1, 2, 3). I suggest we try
> . . . and. . . .

> A way of exploiting the . . . component has occurred
> to me—why don't we use it in our . . . system? It meets
> all the requirements of. . . .

Do you want to describe a minor technical problem you are
having, in hope of getting some help?

We have been delayed in the . . . development because of our trouble in locating a suitable diode. We
have already tried. . . . What we need is a diode that
will. . . .

Poor quality of staining is again bugging us in Project
X. The problem is that. . . .

Do you have some intelligence to report?

Decision to award a contract for the . . . is being held
up because Navy procurement does not like our delivery
schedule. A conference on this problem will be held tomorrow at 9 A.M. in my office to see what we can do
about it. The specific complaint is. . . .

Our . . . paint has been in use for almost a year now,
and customers are reporting that it does not stand up
well. Attached are copies of typical complaints. I would
like to have your comments.

Do you want to request new equipment?

A new . . . is needed for the Materials Laboratory.
We have been spending more money repairing our present one than a new one would cost. The best model to
buy is. . . .

A . . . is needed for precise measurements of . . . in
the Plastic Coating Department. Commercial models
available range from . . . to . . . in cost. Attached are
brochures describing them all. The . . . seems best for
our needs but is close to the top in price. Can I have your
recommendations?

It is apparent that these memos are called informal because
they are informal. The intent is to write as you speak. What
justification can there be for writing any other way?

It is equally apparent that these memos get right to the
point, as good technical writing always does.

It is further evident that these memos are kept as short as
possible. Readers do not want to be bothered with unnecessary coverage of information that is not immediately useful.

Therefore, as you sit down to write such a memo, ask yourself why you are writing, what you want to tell your reader,

what you want him to do. When you have all this in mind, your memo will almost write itself.

You may also set a style for your company.

CONFERENCE NOTES

It is far from easy to act as participant in a conference and at the same time to serve as secretary responsible for taking minutes, organizing and issuing them as a document acceptable as a true account by all present at the conference. Yet engineers and scientists often face this job.

One way to help matters, even though it is far from an ideal solution, is to have someone at the conference whose only function is to take minutes. This person has to be more than a stenographer, because stenographic notes on a conference lasting hours are too voluminous to be useful even when the stenographer is skillful enough to record everything going on. Using a tape recorder is not adequate, because the tape has to be transcribed and the stenographer is in trouble on that score. If a person is available who has sufficient technical knowledge to be able to extract the sense of a meeting as it goes on, he may also be a person who should participate actively in the conference.

The chairman of a conference can help the secretary considerably by issuing an agenda before the conference gets under way, preferably at the time the conference is announced. This will not only keep the meeting orderly, itself a help to the secretary, but it will also provide an outline of the notes the secretary must take. Before the meeting concludes, it is the responsibility of the chairman to restate the agreements of the conference and the status of all disagreements. This will help the secretary check his minutes and give the participants a final chance to clarify statements they have made.

The form the conference notes will eventually take is dictated to a great extent by the topics covered in the conference, their relative importance, and their interrelationships. But there are other considerations independent of the events of any particular conference.

Whatever the reader of conference notes wants to know first is what he must be told first. Surely he first wants a quick summary of the major conclusions of the conference—an abstract.

The most important conclusions may have been decisions to delay rather than to act, to reconsider rather than to decide, to call for more information, for another conference, for help previously unavailable:

> Because complete data are still not available in regard to production costs of the . . . , decision was delayed at a conference held on February 10, 1963, on whether to go ahead with development.

In a happier vein, the most important conclusions may have called for action:

> Development funds of $ . . . for the . . . were voted on February 10, 1963, by the New Products Committee. It appears that production runs of . . . units will enable us to sell at a unit cost of $ . . . .

With either of these openers, a reader knows the gist of the conference notes he has in hand. He does not have to dig out the information he wants most. As he reads the rest of the notes—if he does—he will be looking for details concerning the decision or the delay.

In general, conferences produce several items of major interest to readers. Therefore, the opening sentences cited would be followed by others, each mentioning briefly one result of the conference. These sentences are effectively cast as individual paragraphs, and they are arranged in descending order of importance, with related matters grouped together. If the major findings are numerous, it is advantageous to organize the opening of conference notes in this manner:

> Because complete data are not available in regard to production cost of the . . . , decision was delayed at a conference held on February 10, 1963, on whether to go ahead with development. Three major items remain to be studied:
>
>   1. . . .
>   2. . . .
>   3. . . .

Each number will be followed by several sentences, all on the same topic. In such a case, it will not be necessary to limit each paragraph to one sentence.

After openings such as these, the reader is prepared for the full discussion of each of the items to be developed in the conference notes. In setting up the body of the notes, the writer will follow the structure set in the opening section, using each of three numbered headings, for example, as a major section heading, with each sentence after it reduced to a subheading.

As he writes, he must take pains to cite all important discussion under each of the items. In reporting an individual comment, he should cite its source. Contributions of dissident members must be given as faithfully as those from participants holding the prevailing opinion.

Somewhere in the conference notes, the writer must list the names of all who took part. One good place for this information is in a footnote at the bottom of the first page of the published notes. Many readers know that the facts of life in their particular department or company dictate that unless Mr. Soandso is present at a meeting, no decision of import can be reached, or unless Suchandsuch Department is represented there is no use even bothering to read the conference notes.

It is clear that the secretary at a conference has a difficult task. If he is not careful, he will have his nose wedged so firmly in his notebook that he will not be able to follow the conference at all. If he does not give some thought to the final form of his notes, he will end up with a deadly blow-by-blow account, accurate but impossible to use. He will bore and confuse his readers even more, perhaps, than the meeting may have bored and confused those present.

Armed with the agenda of the conference and briefed in advance by the chairman who usually knows the various points of view that are going to be argued, the secretary can have his notebook arranged topically. As the conference moves along, he can move with it through his notebook. As the conference shifts back and forth between topics, he can do the same in his notebook. When he is finished, he has a notebook that is organized and easy to use. He is ready to talk his notes over with the chairman and get down to writing.

CHAPTER 4

THE TECHNICAL MEMORANDUM

The technical memorandum is a report of research or development. It is intended to disclose fully the result of the effort. No attempt is made to convince or persuade. If ever a scientist or engineer has the chance to write simply and clearly, it is in the technical memorandum. He has information, and his readers want to be informed. Yet, as happens with so much technical writing, his readers are not all equally capable of understanding everything the writer has to say. Some discussion is in order, therefore, of who his readers are and what their special needs may be. The needs of these readers will determine how the technical memorandum—memo for short—is to be organized and written.

PURPOSES OF THE MEMO

A memo may be commissioned by a customer or by the laboratory employing the writer. In either case the readers comprise a large variety of people with many different needs. If anything, the variety of readers within a company is the larger. Let us restrict our discussion to the needs of this larger group, then, knowing that everything said about it will also serve the other.

The memo establishes a record of findings that is intended to inform some readers and instruct others.

In establishing the record of his findings, the writer is putting the logical finishing touches to a project he has carried through. After all, if he does not record his findings, he might just as well not have performed his study—unless he is working for his own benefit only, an unlikely condition in these times. Indeed, at the Bell Telephone Laboratories, it is often said that the only product of the Laboratories is the paper that memos are written on.

There is another reason for thinking of memo writing as the necessary final step in a study. Many scientists and engineers find it helpful to step back and look at their work from a distance, as it were, asking, "What has been accomplished? What can I now say that I was not able to say before this work was begun?"

When we think of how much time and effort go into a study, it is understandable that a man must take a long hard look at his work before he can tell others about it intelligently. The planning and the writing of the memo are that long hard look.

Yet the memo is not written for the writer, but for the readers. If we identify those readers, it will become clear that the memo must instruct as well as inform.

- Some readers have professional training equivalent to the writer's.
- Some readers work in the same field as the writer, others in related fields, still others in apparently unrelated fields.
- Some readers carry on research and development, some assist in research, some supervise it, some supervise supervisors.
- Some readers read to unknot a problem occupying them at the moment, others to keep up with what is going on within the company, others to decide whether such work should be dropped, continued, or expanded.
- Some readers read to see whether theoretical work is needed to understand the questions raised by a study, some to see whether some practical applications of the study can be made.
- Some readers are not yet employed by the company at the time of writing, but will be in a month, a year, five years.

Does it seem impossible to satisfy this audience? The key to a satisfactory solution is found in how you structure your report and how you write each section of it.

ORGANIZATION OF THE MEMO

The memo has six parts:

- title
- abstract

- introduction
- body
- summary and conclusions
- appendix

Bearing in mind the different demands of the readers, let us examine the function, organization, and writing of each of these six parts.

The title

How much damage is done by titles that suggest far broader scope for a memo than the memo actually delivers? How about the title *Morphologic Algorithms?* Surely the memo that bears this one is an all-inclusive treatise, comparable to *War and Peace* and *The Decline and Fall of the Roman Empire.* Pity the reader who picks up this memo hoping to learn all about the subject and finds that the paper should really have been entitled *Computer Techniques Employing Morphologic Algorithms.*

How much damage is done by titles that are too narrow in the sense that they touch on only a corner of the topic actually treated in the paper? Worse yet, some papers bear titles covering material never discussed at all! When *Forest Ecology* turns out to be a study of the effects of ant infestations on field equipment, the reader is justified in committing whatever acts of violence he enjoys.

Have you developed a product? Do you have a new method? Have you field-tested, bench-tested, chemically or mechanically analyzed? Do you have a circuit for, a computer program for, a laboratory test for? Have you subjected a product to preliminary study, feasibility study, commercial-use study? Have you designed something, determined its interactions with, or advantages over, something else?

Say just what you have done—in the title! But don't say more. The title has the purpose of attracting all those concerned with your subject, no matter what the degree of that concern. It must not attract anyone who is not concerned with your subject.

Memo titles often run to much greater length than book or magazine titles. Perhaps this is because the two have different goals—don't publishers usually want to sell to everybody? Memo titles two and three lines long should not be encouraged, but it is far better to write a long title than to cut a title to the point where it is short, snappy, and useless. Occasionally a title will be ludicrously long only because the writer has not looked hard enough at what the title really says. This happened in the case of a paper that should have been entitled *Problems of Telephone Booth Ventilation.* The actual title will not be mentioned here to save the writer embarrassment, but it ran almost four complete lines. Making four lines of what can be said in five words is not easy, but it can be done.

It is good practice to begin writing a memo with only a working title in hand. A good title may come to mind during the hours spent writing the memo. More than likely, several potential titles will suggest themselves. Each one should be written down. When the time comes to select the title your memo finally will bear, you will have the opportunity to consider the various possibilities. Chances are that one of your original thoughts will fill the bill well. If you are uncertain that you have the best title, you can seek help from one of your associates. You may be too close to your paper to see it clearly.

There is one other consideration in selecting a title: memos, like practically everything else in this world, end up in files. In most cases files are arranged alphabetically by titles. If a title is going to help the poor souls going through a file, nothing is achieved by beginning with the articles *A, An,* or *The.* These articles are properly omitted in filing. By the same token, words such as *analysis, development,* and *feasibility* also make for difficulty when placed first in a title. This means, then, that the two-part title is best for filing purposes: *Morphologic Algorithms—A Technique for Circuit Design.* If filing is done alphabetically within subject classifications, the two-part title can be avoided.

The abstract

The memo abstract gives your reader a chance to see what the memo is going to say about the subject mentioned in your title. Unlike the nonabstract, the abstract is not a restatement of your title, but an amplification of it, a brief statement of the essence of the entire paper that follows. The abstract has already been described in a chapter of its own, but some further discussion is in order here.

Brevity is important. Because the abstract is attached to the memo, one paragraph will suffice.

Proper selection of subject matter is also important. Many of those who pick up your memo will satisfy their needs by

A few years later, architect R. Jackson Smith converted the buzzword chart to an "Architectural Innovator System," or "1,000 Guideposts to the Patter-Oriented-Process for Sloganized Architecture." Herewith, Smith's P-O-P system:

Directions

1. Choose any three-digit number.
2. Find the word corresponding to each digit.
3. The three words should indicate an architectural innovation.

Column 1	Column 2	Column 3
1. architectural	1. engineered	1. designs
2. team	2. articulated	2. planning
3. computer	3. coordinated	3. projects
4. field	4. correlated	4. environment
5. behavior	5. modulated	5. schematics
6. group	6. directed	6. processes
7. resource	7. integrated	7. systems
8. realistic	8. controlled	8. concepts
9. value	9. oriented	9. structures
0. creative	0. programmed	0. assemblies

ese, and similar aberrations of the English language. The original Wordsmanship method, also known as the Systematic Buzz Phrase Projector, was developed by a Washington, D.C., civil servant several years ago. Employing a lexicon of thirty carefully chosen "buzzwords," it looked like this:

Column 1	Column 2	Column 3
0. integrated	0. management	0. options
1. total	1. organizational	1. flexibility
2. systematized	2. monitored	2. capability
3. parallel	3. reciprocal	3. mobility
4. functional	4. digital	4. programming
5. responsive	5. logistical	5. concept
6. optional	6. transitional	6. time-phase
7. synchronized	7. incremental	7. projection
8. compatible	8. third-generation	8. hardware
9. balanced	9. policy	9. contingency

The procedure is simple. Think of any three-digit number, then select the corresponding buzzword from each column. For instance, number 257 produces "systematized logistical projection," a phrase that can be dropped into virtually any report with that ring of decisive, knowledgeable authority so necessary to successful technical writing.

HOW TO WIN AT WORDSMANSHIP

After years of hacking through etymological thickets at the U.S. Public Health Service, a 63-year-old official named Philip Broughton hit upon a sure-fire method for converting frustration into fulfillment (jargonwise). Euphemistically called the Systematic Buzz Phrase Projector, Broughton's system employs a lexicon of 30 carefully chosen "buzzwords":

COLUMN 1		COLUMN 2		COLUMN 3	
0.	integrated	0.	management	0.	options
1.	total	1.	organizational	1.	flexibility
2.	systematized	2.	monitored	2.	capability
3.	parallel	3.	reciprocal	3.	mobility
4.	functional	4.	digital	4.	programing
5.	responsive	5.	logistical	5.	concept
6.	optional	6.	transitional	6.	time-phase
7.	synchronized	7.	incremental	7.	projection
8.	compatible	8.	third-generation	8.	hardware
9.	balanced	9.	policy	9.	contingency

The procedure is simple. Think of any three-digit number, then select the corresponding buzzword from each column. For instance, number 257 produces "systematized logistical projection," a phrase that can be dropped into virtually any report with that ring of decisive, knowledgeable authority. "No one will have the remotest idea of what you're talking about," says Broughton, "but the important thing is that they're not about to admit it."

reading only the abstract. They may have misconstrued the title—through no fault of yours, of course. They may be in that class of Very Important Readers who want to have a quick look at every memo turned out in order to stay abreast of new developments. They may be memo shoppers who want a little more than a title gives before deciding whether to read a paper. Whatever the reader's reasons for reading your abstract, he wants certain important information and he wants it fast. He wants an abstract, not a nonabstract.

- What has been accomplished?
- What are the outstanding facts in the memo?
- Is any action recommended as a result of your study, or are there any applications of your findings to point out?

In putting your abstract together, bear in mind that your reader does not know as much of the subject as you. When he has finished the entire memo, he will know a great deal, but he has not yet read your memo. The chances are that at the moment you complete a memo, you are the world's leading expert in the subject. This fact limits the depth of the information you can give in the abstract.

Thus, the abstract is the place for interpretation—for qualitative rather than quantitative information. Of course, some readers will know that a voltage of a specified amount is impressively small, but many others will not. Instead of telling that you used that voltage and leave it to the reader to appreciate the significance, state that you used "as small a voltage as . . ." or that you used "a very low voltage." Instead of telling that you were able to produce a material with a purity of a certain amount, state that "exceptionally high purity was achieved in . . ." or "exceptionally high (. . .%) purity was achieved."

The abstract is the place for well-established, commonly understood terminology. It is not the place for new coinages, arcane jargon, or foreign expressions. Every reader who bothers to pick up your paper because he has been attracted by your title should have a fair chance of understanding what your abstract is talking about. If you feel a brand-new term is necessary in your abstract, do your reader the courtesy of defining it as you use it.

If your abstract has done its job, your reader has spent his time well. He either knows enough to realize that he should not go on with the rest of your memo, or he knows that he must go on because your abstract has made it clear that he will benefit from further reading.

The introduction

Your reader has decided to turn the page. Except in extremely brief memos, an introduction is what he comes upon. What does he find in it?

The introduction to a memo supplies several types of information, but they all work toward one goal: *the introduction helps the reader understand the rest of the paper.*

What kind of help do readers need? The answer to this question must be given by every memo writer each time he writes. It varies from "none whatsoever" to "so much that the introduction will probably be longer than the rest of the memo." The reasoning behind the answer is based on the writer's knowledge of his potential readers and of what his memo is going to present.

Consider a memo reporting the development of an improved technique for some phase of a routine laboratory procedure. Are the readers familiar with existing techniques? It would be a terrible waste of time to give a history or description of earlier techniques in the introduction to that memo.

Consider a memo reporting a radical concept in a field that has only recently come to the fore but will have tremendous impact on an entire technology. No more than a handful of readers will know anything about the subject, and the circulation of the memo will be wide. The introduction to this memo will have to do a great deal for its readers.

A clear statement of the problem the study set out to solve is a good opening section of an introduction. The reason for undertaking the study and the limits of the study also are helpful to the reader who will go on to read the rest of the memo. Such statements can often be made in single sentences. One paragraph is the outside limit in most cases.

Many other kinds of information help readers understand the body of a memo. Here are some questions that can be answered in an introduction:

- What other attempts have been made to achieve what is being reported in this memo?
- What contributions by earlier workers have led to the present development?

- What important sources are available for further study by the interested reader?
- What shortcomings in existing knowledge did the study try to overcome?
- Is there specialized terminology in the body of the memo that can be defined now in order to help the reader?
- Is there theory that should be explained here because the body of the memo leans so heavily on it?
- Is the present study part of a larger project the reader should be acquainted with in order to grasp the memo fully?

It is obvious that this entire list of questions need not be answered in all introductions, but if the writer feels that any of this information would help his readers, he should supply it.

There are ways of limiting the extent of the answers to some of these questions. If information exists in standard textbook sources, there is no reason to include it in a memo. If earlier memos have covered the ground adequately, the writer does not go over it in detail, but supplies only a quick summary for his reader's convenience and then cites the appropriate sources. The memo is not a dissertation; there is no need for exhaustive historical treatment.

One final help the introduction gives is to supply an informal table of contents of the remainder of the memo, since only memos of great length justify inclusion of a formal table of contents. The informal table of contents is a paragraph, naming major sections of the rest of the memo in the order in which they will appear. Important illustrations in the body and the subjects treated in appendices are also mentioned to help guide the reader.

Considering the dual functions of the memo—to inform and to instruct—the introduction to a memo is primarily a teaching section. It is an important part of the memo.

The body

The memo body describes procedures followed and materials used to achieve the results of your study. It details tests and test results. It presents the analysis of the data collected. The body is the foundation of the memo.

Its organization is simple. What must the reader be told

first so that he can understand what follows? What must he be told next, and next, and next?

In a typical memo, the first item treated in the body is the experimental set-up. "Equipment and Material Used" is the first section, "Procedures" followed the second.

A sketch can frequently carry much of the burden of description. It gives the reader a better picture than words can of what the equipment actually looked like and how it was arranged. This is most striking when the experimental set-up is primarily a circuit or some chemical glass. On the other hand, when it comes to describing procedures, pictures are of little help. If a procedure is given step by step, the writer does best to rely on a tabulated presentation. Readers grasp this kind of arrangement easily, and if a reader attempts to duplicate the procedures, he will be able to work most easily from a tabulation.

The amount of detail in the description of equipment and materials is a decision the writer makes based on his knowledge of his readers. Careful attention to proper nomenclature is always necessary, but in memos intended for new personnel or technicians, the writer must take special pains to be sure that the terms he uses will be understood.

This problem is even more acute when procedures are described. New personnel frequently are penalized because they are not familiar with local terminology. Again, they suffer when a writer lumps several stages of a process together as though they were one. Of course, the writer has been performing a particular laboratory procedure for so many years that what is really an involved matter seems to be an easy routine. The new man is not that lucky. If the writer is absolutely sure that his memo will never be read by anyone except experienced personnel, the presentation can be brief, and a few words may replace many lines of description.

Once the reader knows how you went about collecting data, he is ready to be told what you found. In presenting "Data Collected" or "Findings," the writer must first select a form for the presentation. Will a table be best, or a curve, or some paragraphs of text? The reader will prefer tables and curves. Yet the solution usually turns out to be a combination of both of these *and* words.

The important considerations in presenting data are completeness, accuracy, and clarity. Is everything there? Are units spelled out? Is the range of accuracy of readings accurately specified? If your conclusions are to stand close inspection, the reader must have all the information he needs, and he must be able to grasp it.

After the presentation of data, there is usually a section devoted to "Analysis of Data." Your reader has to know how you handled your data in arriving at your conclusions. This section is straightforward, presenting few problems to the writer. The only serious mistakes that can be made are supplying too little information or omitting it entirely when needed. If unusual mathematical procedures were used in handling the data, the reader must be given enough information to understand your procedures. This does not mean explaining established routines, nor does it mean introducing detailed proofs if they are available in standard references. When the necessary explanation is not readily available elsewhere, it is often better to refer the reader to an appendix rather than to clutter the memo body with pages of formulas.

A few words might be said here about the writing style of the memo body. The rule throughout is to be as simple and clear as possible, no matter who your readers are. New engineers and scientists are often guilty of assuming that all their readers are as bright and well-informed as they are. Their memos generally turn out to be suitable reading only for themselves. Remember that your reader would rather have the prerogative of skipping some material than worrying about why he cannot understand your memo.

- Outline the body before beginning to write it and write each section on separate sheets of paper so that you can rearrange them easily if you discover that your original order of presentation is not the best possible.
- Decide on the proper name of each component, material, device, technique, and tool you discuss and stay with that name. Readers find it upsetting when they have to perform mental gymnastics to work out synonyms the writer thought clever. A transistor may be a semiconductor, but why not call it a transistor every time?
- Use the simple past tense in describing things you did in the laboratory.
- Use the simple past tense in referring to the work of others.
- Avoid sentences that go on and on and on.
- In presenting a proof, don't commit the sin of skipping necessary steps because you are such a hot-shot mathematician.

The classic example of step-skipping came in a lecture given some years ago by Professor Norbert Wiener before a meeting of statisticians. Most of the people in attendance were first-year statistics students, but the evening was catastrophic even for the knowledgeable.

Wiener opened his talk by puffing on his cigar for some minutes after being introduced. His brow furrowed. He paced back and forth. The tension grew. Finally he wrote on the blackboard:

$$a = b + c$$

Everyone sat straight in his chair. This wasn't going to be so bad. Wiener puffed some more and paced back and forth again as he stared at what he had written. Several times he approached the blackboard, apparently ready to write again. Each time he backed off. Then he said his first words of the evening—those glorious words heard so often: "From which it is evident that——" . . . and he wrote one grand equation that filled the blackboard. The integrals and sigmas were everywhere. Disaster.

Such behavior, even a mild approximation of such behavior, cannot be countenanced in a memo.

Summary and conclusions

Having read why and how a set of findings has emerged, and what those findings were and how they were treated, the reader is ready for a restatement of the most important findings and the conclusions drawn from them. He should not be surprised by anything he reads in the "Summary and Conclusions" section of a report, for the abstract has already tipped the writer's hand. The body presented all data collected. If much that is new appears as summary and conclusions, the rest of the memo has not done its job. Is this section, then, nothing more than unnecessary repetition of what has gone before? By no means.

What was given in qualitative terms in the abstract is spelled out here in precise numbers. What was given in the data section of the body appears here only if it is worth additional emphasis.

The "Summary and Conclusions" section is a brief statement of the most important findings presented in the memo. It is in a form suitable for those readers who would like to preserve essential notes on the achievement described without bothering to save what may be too bulky a document for their personal files. All they need do is pull one sheet out of the memo, the sheet containing the summary and the conclusions.

Each sentence is a paragraph, reflecting the importance of the statement. Of course, some thoughts may be so complex that they cannot be expressed in a single sentence of less than homicidal length. But the writer does well to examine any thought that seems so complex. Perhaps its substance is not all it appeared to be at first glance.

Each sentence is written with all the force and self-assurance of an ex cathedra pronouncement. There is no need to support it—the body of the memo has done that already. No references to figures are necessary—the reader has already studied them. Just make your point and stop.

> "The estimated cost of producing . . . by the . . . process is. . . ."
> "Distillation is unsuitable for. . . ."
> "Further study is needed before . . . can be installed."

Some writers choose to separate summary statements from conclusions. This practice is acceptable in a memo that must otherwise have a lengthy "Summary and Conclusions" section. In fact, however, most memos do not contain both summary and conclusion material. Even if they do, there generally is not sufficient material of both kinds to support separate sections. In combining both kinds of information in the single section, the writer must make clear to his reader which of his statements are summary, which conclusion:

> "A noise figure of . . . was obtained through . . ." is *summary*.
> "The noise figure of . . . obtained through . . . is satisfactory for use in . . ." is *conclusion*.

In some companies the laboratory-practices manual makes a distinction between conclusions and inferences. A few even go so far as to differentiate inferences from recommendations. And the writer is expected to write something under each heading in his memos. This smacks of Parkinson's Law. No matter how many sections a report form calls for, writers will manage to write something for each. But is this writing necessary?

A final note on this penultimate section of the memo. In many laboratories the practice is to include a short statement of recognition of the help given the writer during those dark hours in the laboratory. This statement is often saved until the very end of the "Summary and Conclusions." Acknowledgments are a noble rite and one to which authors are almost universally addicted (see the dedicatory page of this work). It must be pointed out, however, that scientists and engineers have been known to lose their professional aplomb and grow maudlin in heaping praise on an immediate superior. Enough said.

The appendix

Like any other appendix, not much that is helpful can be said about the appendix that appears at the end of a memo. It spares the reader what can be irksome side journeys while he is going through a memo in pursuit of information and ideas. It serves as the repository of information the reader should have to understand the work fully, information withheld until the final pages because it would take too much of the reader's time as he works through the body or because it may be highly specialized and not of interest to many. The appendix can also present an obiter dictum not too closely related to the main stream of the memo. Warning: don't heap appendix on appendix unnecessarily. Consider issuing a separate memo instead. Lots of worthwhile information lies buried, unread, in many an appendix.

SOME ADDITIONAL CONSIDERATIONS

Writers often wonder where to place illustrations in a memo. A typist hates to be asked to insert them in a text where they are mentioned. Illustrations saved for the end of a memo mean a happy typist but not a happy reader. What should be done?

In most cases the typist wins. After all, where can you get all the typists you need these days? But why should the reader lose? We have worked so hard in this chapter arranging material so that he can read easily and profitably.

Short of insisting that he have his way—unthinkable—the writer can try to get the typist to use pull-out sheets for illustrations, so that the reader can consult a figure at the same time that he has his textual page before him. The typist can collect all these figures at the end of the memo, and the reader is assuaged.

Responsibility for editing and proofreading a memo is entirely the writer's. No matter how tiresome the chore, no one else can take it over. Yet there is much to be gained from entering into a co-operative project with other writers, in which Castor will read memos written by Pollux, and vice versa. Even so, the writer will give his own memos thorough editing and, after the typist has completed her first draft, thorough proof-reading. It is silly to spend hours writing a memo that represents months of work and then quit before the finishing touches have been put on it.

In the same vein, many an engineer and scientist mistakenly delegates responsibility for correct spelling to the departmental typist. This is too often a case of confidence misplaced. Everything in a memo is the responsibility of the writer.

THE JOURNAL ARTICLE

There is an apt expression used by baseball players to describe the failure of a player or—what is worse—an umpire to perform well when the going gets rough. They call that failure "choking up."

All too often scientists and engineers exhibit a similar reaction when faced by the need to produce an article for publication in a journal. They may have successfully written many memos of all kinds, may even have reached the point at which writing is almost something to be enjoyed. Yet, when they find themselves in the position of having to turn out a journal article, they act like beginners.

What is there about "publication" that causes this to happen? Are the readers of journal articles so different from readers of memos circulated within a company? Are the words they understand so different? Is the standard of writing higher? Is the journal editor an ogre?

With only minor exceptions, the form of a journal article is substantially the same as that of a memo. Requirements of style, tone, and language are no more rigorous. Clarity, completeness, and conciseness must be striven for.

Knowing what you have to say, planning how best to say it, and then writing it for people rather than for posterity will see you through. There is no reason why you should abandon straightforward and informative writing for a stilted, "Look at me—I'm eminent" style. There is no reason why you should "choke up."

THE AUDIENCE

The writer of a journal article generally is familiar with the journal he is writing for. He can therefore assess accurately the degee of technical competence the journal assumes its readers to possess. In the event he is writing for a journal he does not know well, he must read through a few back issues in order to get a picture of the typical reader. After all, if he is to write for that person, his choice of language, his assumptions as to what the reader knows, and his article format must be based on that reader's needs.

Journals make some effort to acquaint writers with their requirements by sending out a pamphlet of advice for writers. In some cases this pamphlet is a sheet or two of mimeographed material so vaguely written that it is next to useless. It may, for example, be given over entirely to matters of physical appearance of the manuscript. In other cases, however, the writer receives a great deal of help. Whatever the degree of helpfulness or the form of the pamphlet, the writer should study it carefully and do his best to comply with the editor's requests.

When specific information is not available as to the nature of the audience for an article, the writer may fall back on a picture of his reader as a person who has had the same academic training as the writer, the same amount of professional experience, and the same interest in the subject of the article. This hypothetical reader stops short of being the writer himself in but one consideration: the reader has not performed the study the writer is reporting.

If you will think of yourself on the day when you first began to consider the specific problem you have solved, or partially solved, you may usually assume safely that you have a valid image of your reader. Do you recall the amount of digging you had to do in order to get started toward a solution? Do you remember the review of fundamental information you gave yourself? Do you remember the false starts and errors you made? Do you remember the technical assistance you needed in order to carry out the study? Your reader has not been at your side throughout the weeks, months, or years since the day on which you began to work. You must not expect him to know, therefore, more than you knew at the start of the project.

It is these considerations that underlie development of a suitable article.

ORGANIZATION

Like the technical memo, the journal article consists of title, abstract, introduction, body, and conclusions. Unlike the memo, it does not usually contain an appendix, because of the

limited space available in journals. The requirements for all
the sections of the technical memo described in Chapter 4
apply equally well to the sections of the article—with one
exception, the introduction.

The introduction to a journal article has two functions:
it describes the problem attacked in the study and it acquaints
the reader with other important work performed in the subject
field.

If the reader is to be able to read an article easily, he must
be told the goals of the work performed. In some introductions, these goals are described at the outset. In most cases,
however, the abstract has satisfied the reader's initial curiosity,
and it is more logical to provide the historical information first
—beginning with the earliest pertinent work in the field—and
lead up to work being reported. In one sense, then, the
introduction may be thought of as a historical treatment of
the efforts and the problems of a technical field that begins
as far back as necessary to get the reader aboard and builds
up rapidly to the day when the reported work began.

This historical account is not as complete, nor should it be,
as the account found in the typical dissertation or thesis. There
is not room for such completeness in a journal. But it must
touch on the highlights of the efforts that preceded the work
described in the article, and it must culminate in a clear statement of the problem undertaken by the writer.

A common fault of writers attempting this job is to supply
too much material. The reader is surely interested in your
account, but there is no point in snowing him under with detail
or with so many references to the literature that he becomes
enmeshed in a bibliographic exercise that smacks of the cult
of personality. A pox on name dropping! The reader assumes
that the author is well read. Sources should be limited to the
most outstanding contributions, sources the reader will find
worthwhile should he decide to read further. To help a reader
who wants to find all the sources in a field, one general reference will do most of the job for him.

In citing a source, be sure to supply enough bibliographic
information so that your reader can find his way easily to the
source. (Detailed discussion of citation forms is provided in
Chapter 15.)

A final word of caution. Some introductions to journal
articles adopt a tone that is completely unbecoming:

Greene made the extraordinary observation that . . .

Smith fortuitously . . .

Jones chanced . . .

Brown, in a happy stroke of serendipity, found . . .

The introduction is concerned with presenting information that will help your reader understand your article and perhaps go on with similar work. Why not stick to the facts and leave the editorializing to others?

NOTETAKING IN LIBRARY RESEARCH

The 5x8 index card is an effective means of recording notes on library research. A card of this size is easily managed and large enough for even the most elaborate notes.

Historical material to be included in an introduction generally falls into not more than two or three subject headings. For example, there may be *theoretical background* for the study being reported, and *technological contributions.* Thus the researcher will have little difficulty in organizing his effort. But there are other practical considerations in notetaking that are worthy of attention.

When making notes of material dealing with any subject heading, three procedures should be followed. First, if the material is to be quoted, care should be taken to record it exactly as it was written. Secondly, citations will be necessary for footnotes, bibliography, or both, so complete information should be placed on each 5x8 card. Finally, accurate subject headings should be used, so that the subsequent arrangement of notes will be easy.

In taking the notes, the researcher either abstracts or quotes directly. A good guide for deciding which of these methods to use is: if the author has stated the concept as simply as possible, or has stated it in such a manner that abstracting would destroy the impressive stylistic quality of the original, then the material should be quoted. If abstracting takes fewer words and retains the main idea, then the material should be abstracted.

Most researchers take more notes than can possibly be used in a paper. Think before you write. Remember the length of the journal article you are planning and the importance of its introduction. Ask yourself, "Approximately how long will this section of the paper be, and how much support do I need to establish the validity of the point I am trying to make?"

Take only one note on a card, even if it is only one sentence. Taking more makes almost impossible the effective arrangement of cards according to topics and the final form of the work. Limited to one note each, the cards can be shuffled and reshuffled as the final form of the introduction evolves. Moreover, with one note to a card, there can be no mistakes in footnotes.

The simplest means of using cards effectively is to put the subject heading at the top of the card and, directly underneath, author, date, and title of the book, publisher, place of publication, and the page on which the material appears. For an entry based on a journal article, the subject heading at the top of the card should be followed by author, date of publication, article title, journal, volume number, and pages on which the material appears.

With these entries complete, the 5x8 cards are working substitutes for the original sources, and the writer will not have to leave his desk during the outlining and writing stages for a time-consuming trip to the library.

If the writer is given his choice of where to place his footnotes in the finished article, he should give greatest consideration to his readers' needs. Clearly, a reader wants to be able to see a footnote on the same page to which it refers. In the case of a footnote that is not a reference to the literature, but an important comment outside the direct line of discussion, the readers' needs are paramount. But the writer is not always free to choose. Most journals are unwilling to print footnotes and a bibliography as well. In addition, they prefer not to pay the additional costs incurred in setting up footnotes at the bottom of a page. In such cases the writer must settle for collecting his bibliographic footnotes at the very end of his article. Such comments as would ordinarily appear at the bottom of a page of text must be incorporated—however unsatisfactorily —in the body of the text. Parentheses must be used to enclose this material.

The entries in a bibliography for a journal article are most often arranged in the order of citation within the article, because they must perform the dual functions of footnote and complete entry. The best guide for form is the journal for which the article is intended. Great differences appear in practice, and it would be next to useless to offer model forms here. Chapter 15 presents examples of the best modern usage.

EDITING AN ARTICLE

Because an article writer and his editor generally are miles apart—geographically, that is—there is little opportunity for the kind of review that some scientists and engineers are used to receiving from a supervisor. (This separation can also be a gap in knowledge, we know. Witness recurrent hoaxes perpetrated by frolicsome article writers whose contributions are entirely imaginary.) In most cases journal articles are read by one or more anonymous critics who concern themselves solely with the technical content of an article rather than with its written quality. The journal editor himself makes few changes in a text beyond correcting obvious flaws. The writer, therefore, has great responsibility for editing his text.

He may choose to have the article read before submission by one or more of his associates, and this practice is extremely helpful if they are willing to be candid. In most companies articles have to be submitted for review before they can be released for outside publication. Prevention of disclosure of proprietary information is often the primary concern of this kind of review, however, and the writer is not helped to the degree he requires. But no matter what kind of help the writer seeks and is given, on him must rest the heaviest burden for seeing that his article turns out to be as good as it can be.

After the first draft is complete, an article should be given several thorough readings to see that it meets the requirements of its readers. At least four different readings are required, and they should be spaced out so that the unpleasant task of editing is done properly.

¶ The first reading is for *content* and *form*:

- Does the title state as clearly as possible the specific subject of the article so that potential readers will not overlook it? Does it promise just what the article delivers—no more and no less?
- Does the abstract come to the point at once, telling the reader the exact findings? Does it use language that can be understood without referring to the body of the paper?
- Does the introduction give a clear picture of important developments prior to your study? Is the statement of the problem clear to someone who does not have long hours of experience with the problem?
- Does the body contain all important information? Has

sufficient detail been supplied to enable a reader to duplicate the study? Has space been wasted on information that is readily available in standard sources? Is the arrangement of elements within the body logical?

- Is your conclusion completely justifiable on the basis of information supplied in the body? Is it as complete as it should be? Are guesses, opinions, and surmises clearly labeled?
- Do you supply all necessary illustrations?
- Do you supply any unnecessary illustrations?

¶ The second reading is for questions of *style*:

- Is all terminology correct and standard? Have you supplied definitions of recent coinages?
- Are there any needless synonyms, euphemisms, or ambiguous constructions?
- Are there any unnecessary words, phrases, or sentences?
- Does each paragraph make one important point?
- Does each paragraph flow into the next?
- Are the section headings as descriptive as they can be?
- Have you cast your sentences in the simplest, most direct, most vigorous language possible?

¶ The third reading is for *details*:

- Are references complete and consistent in form?
- Is every word spelled correctly?
- Is all punctuation correct?
- Are figures properly numbered and named?

These three readings are necessary if your article is to be effective. The best way, incidentally, is to read aloud to yourself—your ear will help your eye.

But you are not yet through with your editorial chores. After you have made all changes in the first draft that you are going to make, your typist will have another go at it to put the article in final form. It is well known that each retyping of an article may introduce new errors. After the final draft is complete, therefore, the typist will proofread carefully. When she has finished, you are ready for the fourth reading, a *proofreading* of the article. A good way to do this is to have the typist read aloud from the first draft while you follow in the final copy. This method will easily discover any typist's errors and possibly a few more stylistic errors or errors of fact that escaped notice earlier.

You and you alone have the ultimate responsibility for correctness of your article, and you must not neglect it. Only after you have braved the tedium of this editing can you dare to send off an article.

ORAL PRESENTATION OF A JOURNAL ARTICLE

When we think of all the professional meetings and conventions that are held in so many parts of the world each year, when we think of the thousands of trips that are made by scientists and engineers to take part in these sessions, when we think of the amount of time away from the laboratory that they represent, we must wonder whether all these meetings are justifiable.

Just as there is at least one national society in the United States for every discipline (more if there has been a split among the leaders), there is often at least one branch of these organizations in each region of the country. A national group usually has one or two meetings a year, regional groups several. Thus we have meetings upon meetings, journals upon journals, articles and articles and articles.

While this book is not about techniques of presenting technical information orally, so many technical articles are first presented as talks before meetings of one kind or another that it is appropriate to spend a little time discussing the relationship between the form of a paper and the talk to which it is related. Hopefully, it may even be possible to make a few helpful suggestions for the engineer faced by the specter of having to give a talk.

The best oral presentations of technical information are delivered extemporaneously. The careful preparation that underlies such talks rivals that needed for writing journal articles. Outlining, planning of visual displays, and rehearsals are all needed. When the speaker finally stands before a meeting he is to address, he is thoroughly familiar with his subject, what he is going to say about it, and the order in which he will say it. The only written material in front of him on the lectern is a page of notes, no more than a list of the topics he will cover in the order he will cover them, with clear signals for changes of slides or charts.

The worst oral presentations of technical information—and

the most typical, unfortunately—are read word for word from prepared texts. In most cases these texts are the articles the speakers have written for the journals published by the sponsors of the meetings. The only changes are the addition of some salutation of the members of the audience and a Bennett Cerf-type joke guaranteed to win over the audience.

This is not to say that all those who read papers before scientific meetings give poor performances. Speakers have been known to read aloud so well that an audience can scarcely tell that the talk comes from a script. But such skillful readers are a tiny minority. In general, performances are so poor, and listeners get so little from them, that it might be better to conduct meetings in some new way.

In one such scheme, all papers would be read aloud simultaneously in a single room, with all speakers lined up elbow to elbow. The program would be held early in the morning, and anyone attending the meeting who is not scheduled to give a paper would be able to stay in bed that morning and get some extra sleep. Once all the papers were out of the way—and this would take only an hour or so—everyone could enjoy the rest of the meeting without having to sit still for hours on end, learning nothing and being bored half to death. Parties, dinners, after-dinner speeches, awards, sightseeing, and interviewing for new jobs could go on as usual. If this solution appears too radical, consider whether most of the meetings you have attended in your professional life were any more productive than the type of meeting proposed.

Should this scheme not succeed in winning support from professional societies, there is another alternative. Under this scheme, all papers to be delivered would be prepared in written form prior to the meeting. Copies would then be mailed to all who register for the meeting. Additional copies would be on hand on the big day for distribution to those who had not done their homework. The technical programs would then be devoted to presentations of additional material, visual displays, and question-and-answer periods.

If the various societies insist on maintaining their present practices in conducting meetings, the least they can do is warn against reading papers aloud. The main objection to this practice, aside from the poor oral reading so prevalent, is that papers that can be read well silently seldom can be read aloud with comparable effect. A good talk simply cannot convey all the information a journal article must contain.

The organization of a talk is necessarily different from that of an article. Talks demand repetition and other rhetorical devices an article does not need. A talk is heard once; a paper can be read many times, if necessary. A reader can think through a difficult point without losing his place in an article; in a talk the information flows on and on. Any listener bothering to think about a difficult idea will find that he has lost the thread of the talk.

Prepare your journal article in the manner described earlier in this chapter and in Chapter 15, but know that if you are going to give a talk on the same subject, additional work will be needed.

Work first on the visual material you will use in your talk. Some of it will be identical with exhibits in your article. Then prepare an outline of your talk. When this is done, stand in front of your visual material and give your talk to a dictating machine. Remember that you are rehearsing a talk to be given to people. Talk to them. Bear in mind that they are hearing your talk for the first time and that they will not be able to hear it again. Be sure they can follow you. As you move from point to point, look at your displays to make sure that they are helping your audience understand.

When you have finished, play back what you have said, and *listen*. Keep your eyes on the visual material throughout. This is what your listeners will be doing during most of the time you will be speaking. As you listen, your strengths and weaknesses will become fairly obvious.

Prepare notes on your talk as you listen. Then give your outline a going over and give your talk again. And again. And again. Each time listen to the playback. Finally you will be satisfied, and your outline and visual material will be in final form.

Now you are ready for a dress rehearsal before a group of your associates. If you will encourage them to be frank in their criticism, you will get an audience reaction that is invaluable. You will also have a taste of the kinds of questions your real audience will ask. If your associates like your talk, the audience you are preparing for will probably like it too.

MANAGEMENT OF TECHNICAL WRITING

Since writing is an essential function in research or development, it is difficult to understand why the management of the writing process is so ineffectual in many modern laboratories. The result of such mismanagement is employee discontent, constant scrambling to meet deadlines, poor writing, and nights and week ends of unnecessary and unpleasant work. A research or development effort is incomplete until a clear report of findings has been issued to inform others of the results. A brilliant idea cannot be sold to, or used by, others until a written document has described it. Inevitably, therefore, the scientist and engineer become writers. It might be added that they become, on the average, the world's most highly paid writers. Yet their supervisors treat them inadequately, at least in this expensive role.

A CLIMATE FOR WRITING

Six management responsibilities must be met in order to create a proper climate for writing in today's laboratory. Each contributes an important element to the picture, and all are easy to achieve.

¶ The first responsibility of management is *to provide a place* for the scientist or engineer to do his writing. Of course every laboratory has desks, supplied with all the latest gear—elegant file drawers, telephone, calendar, and the like. And a chair stands in front of each desk. What more can a man ask?

A library is nearby, complete with shelves and files of all pertinent scientific and technical publications. A librarian stands ready to perform literature searches, compile bibliographies, abstract current publications, and list new purchases. Card catalogs and reader's guides are at hand. A photocopy machine is available to save time-consuming transcription. Tables and chairs provide work space. What else is needed?

The one essential for every writer, and one not often provided the technical writer, is *privacy*. The United Nations some years ago set aside a Quiet Room so its delegates could meditate. University libraries have their cubicles to protect thesis writers from interruption. Every author has his own quiet place in which to work. Yet many of our scientific laboratories, resplendent in their shiny new glass and marble elegance, seem to go out of their way to place obstacles in the path of good writing.

To do his job, the writer does not need complete isolation, but when he shares a room with other scientists or engineers, he is subjected to every interruption impinging on every one of his roommates. A telephone ringing anywhere in a common room affects everyone in the room. Let no man send to ask for whom the telephone jangles, it jangles for thee. A visitor for one man becomes a visitor for all the others. A conference between two men in a room distracts everyone else. If all the men in a room are writing, then there can be a thousand men in the room—but all must be writing, not interviewing salesmen, conferring on design problems, making long-distance telephone calls.

In the absence of space for private offices for all members of a professional staff, a special room must be set aside for writing. No telephones. No conferences. No typewriters. One table, one chair for each writer, and a small partition on either side of him. The only interruption that is permitted is a whispered message from a clerk that urgent matters require attention *outside the writing room*.

¶ The second responsibility of management is to see that *each writer is given time* in which to carry out his writing task.

In budgeting time for a project, almost every aspect of the work is given its appropriate allocation—planning, requisition and purchase of materials and equipment, staffing, study, testing, liaison with customers, and the like. As for time needed for writing, it is too often assumed that little men are going to appear each night at the stroke of twelve to pick up the data scattered during the day and weave them into a magnificent report fabric. Even when time is budgeted for writing, when a project schedule runs afoul of the calendar, the needed days and weeks are stolen from writing time. Writing always takes time, and time must always be set aside for writing.

There are many ways in which writing time can be cut down, but it must be recognized that no matter how skillful the writer, he must have time available to complete his task. Often men are assigned new problems to solve before they can record the results of their most recent task. A cook cannot go on and on preparing meals without cleaning up the mess left in the kitchen by each preparation.

¶ The third responsibility of management is to see that *the requirements for each writing task are as clear as they can be before* the writer gets down to work.

A *standard* for all types of reports written in a laboratory is one necessity. What kinds of information should each report contain? What section headings must be included? What physical format is expected? What style of writing is required?

In most cases this can best be handled by two documents: a few pages of *mimeographed instructions* addressed to researchers and the clerical staff, and one of the dozens of published *manuals of writing*. (The second half of this book may do the job!) The mimeographed material should be drafted by an engineer or scientist on the staff of the laboratory, not by the chief of a group whose job it is to turn out instruction manuals and the like. The written reports created by engineers and scientists have their own special requirements, and they will be written by men who are not full-time writers.

Since two documents—mimeographed instructions and a style manual—are called for, the engineer responsible for the first document is cautioned to avoid posing as a Fowler or Perrin. His job is not to comment on the English language, semantics, or writing style. He must limit himself to drafting the requirements of content and format for all types of reports done in his laboratory. The manual of style will do the rest and do it better.

When the first draft of these instructions is complete, it should be read and criticized by a committee of senior staff members. When this group has done its work, and there is agreement on the contents, mimeographing and distribution of the instructions can begin. Of course the finished product will immediately be on its way to a revised edition, but that is a necessary and healthy process. If a second edition is not soon forthcoming, it is only because everyone has filed and forgotten the instructions.

Before a man can get down to writing a report effectively, he must have a *conference with his supervisor*. No matter how explicit and complete the set of instructions issued by the laboratory, no matter how clear the writing manual, each writing task will have special requirements that can only be conveyed face to face. Should a writer be given some particular insight into the audience intended for a report? Are there any so-called political problems—toes that must not be stepped on, points of view to be avoided, and the like? Is certain information to be withheld or revealed in another report? Nothing destroys a writer's spirit quite as completely as being told after he has worked hard on a long and difficult report that he should have written it in some special way and that he now can tear up what he has done and begin all over again.

The most useful conference between supervisor and staff member covers all these points before writing, but it does even more. If a supervisor wants to earn the undying gratitude of one of his men, and teach him a great deal about writing so that he can save much time on all future reports, he will develop the outline of the report with the man before he lets him go. Beginning writers in particular, but even experienced writers, have difficulty in organizing a report properly. The main reason, beyond inexperience, is that a man is often too close to his project to see what he has accomplished. The supervisor is one step removed from the work and can better appreciate the significance of data. He also is more aware than the researcher can possibly be of considerations affecting the report.

The final challenge of the successful conference is to draft the abstract of the report. The ideas and the wording by which they are expressed can be a joint effort, but with a new engineer or scientist the supervisor must be prepared to do this writing himself until his man gets the hang of what is desired in an abstract. A writer going back to his desk armed with an outline, abstract, and notes covering all the special considerations of a report is a writer ready to work.

¶ The fourth responsibility of management is to see that *editing of reports is done honestly and promptly.*

When a report is given to a supervisor for approval, it has already been edited carefully by the writer with the help of his associates and secretary. (Chapter 5 describes this procedure.) The writer feels that the report represents the best work of which he is capable. This pride of authorship is a secret failing of everyone who has ever written a sentence. Even if a man says to his closest friend that he is really a poor writer, down deep inside him he knows that he is destined for immortality beside Tolstoy and a few others. He is surely mistaken, but he must be told so with a perfect combination of tact and vigor, so that he can improve his report and do a better and better job on each report he writes from then on.

- Not a single word must be changed in a report without an accompanying explanation.
- Not a single sentence must be written by the supervisor until the writer himself has had a chance to try again once the problem has been explained to him.

These two warnings do not apply to lapses of syntax, spelling, or other mechanical matters discussed in the style manual. Anything that can be supported by resort to the mimeographed writing instructions is also fair game. But supervisors are hereby warned: be certain that you can find support in these "official sources." There is something truly ludicrous in the supervisor who thinks himself an authority on writing matters when he really knows little. Mrs. Kowalski may have done her job well in teaching you fifth-grade English, but you may not remember all she said. Also, a lot she said then is no longer necessarily true. A man who stands as a judge of writing without the backing of authority in usage eventually ends up with two types of men working for him, those who perpetrate his own linguistic indecencies and those who cannot write at all because they refuse to conform. Every new supervisor of research and development should be required to take a course in writing and editing.

Perhaps the most frequent type of management error in editing is to change a word, phrase, or sentence a man has written simply because "that is not the way I like to write it." If there is no strong reason for changing a man's words, they had best be left alone. And the only strong reason beyond a political one is *clarity* (scientific authenticity is, of course, assumed). The explanation to a writer that his way is unclear should lead first to another attempt by the writer. Only if he fails again should the supervisor rewrite.

Up to now, this discussion of the supervisor's responsibility as editor has been limited to the question of honesty in editing. The other requirement stated was promptness.

A special corner of hell is reserved for those who torment writers by not returning their work on time. How discouraging it is for a writer to sit at his desk day after day, wondering what has happened to a report he turned in. "It's probably in the hands of the laboratory director or the president of the corporation. They want to fire me or make me director of technical publications. They want me to give the paper before a technical society. They have thrown it into the wastepaper basket. Just who does that supervisor think he is? It is hopeless. It is just fine." As he sits and curses his fate, Dr. Miniver Cheevy is not in a mood for scientific discoveries and dramatic advances in the art. As the weeks go by, he becomes less and less optimistic, more and more upset. An out-of-sight report is not out of mind; it only causes trouble.

A report should be returned as quickly as possible after it has been submitted, and many supervisors make it a practice to hold reports no longer than forty-eight hours without making some kind of explanation to the writer. A report that passes inspection without a great deal of blue-penciling can be returned without a conference, but one that requires much editorial comment calls for a conference with the writer.

The emphasis in such discussions should be on ways in which the writer can improve his own report. There is more at stake than the perfection of one document. First of all, there is the relationship between the man and his supervisor. Quality of writing is difficult to agree on, so careful explanation is needed if the two men are to understand one another well. Disregard of the feelings of the writer can lead to permanent damage in the relationship between the two men. Then there are the many reports the writer will turn out in the years ahead. If the conference does its job, errors similar to those made in a given report will not appear in future reports. Finally, there is the training a writer gets in the art of supervision, an art he will be called on one day to practice himself.

In reading a man's report before returning it to him for corrections or rewriting, the supervisor should guard against the mistake of acting solely as a proofreader. He can be thorough in his search for bloopers, but his first responsibility is to try to picture how the actual readers of the report will react to the document. This calls for reading in the same way as those persons will read, and they will not be looking for small errors. Broad impressions are most important at the outset; clarity of presentation comes next. If a supervisor doesn't "get" something easily, then other readers will not either.

¶ The fifth responsibility of management is to *recognize and reward good writing*.

The love of the writer for what he has written has been mentioned before. The importance of recognizing this failing of writers cannot be mentioned too often. While many writers assess too highly the quality of their work, there are plenty of good writers, and every laboratory has its share of them. Good writing reflects clear thinking, and it is natural that jobs that require a high order of intelligence should attract a great number of people who can express themselves well. It is important to recognize this ability to write well for two reasons.

First of all, good writers who are told they are good will generally become better and better. Conversely, good writers who are not told that they are good will feel less confident about their ability and not perform as well as they can.

Secondly, poor writers can benefit from encouragement when they show signs of improvement. Thus it is important for a supervisor who is critical of a report to point out whatever virtues a piece of writing may have. The poor writer will know just what his problems are and will concentrate on solving them, confident that all is not hopeless.

It is good practice in a laboratory to call the attention of all scientists and engineers to an outstanding job of report writing by one of their associates. In one laboratory, there is a "Report of the Month" competition. The laboratory director, *who reads every report written by his staff,* issues a memorandum praising the report or reports he finds outstanding each month—if he finds any. He not only names the report and the writer, but states in a paragraph or two what he found worthwhile in the writing. This same man also writes a sentence or two to many other writers each month, praising or criticizing reports they have done. A New York *Times* editor each Monday circulates a list of good things and bad he found in news stories during the previous week. Showing a staff that its work is read "all the way up the line" is a sure way of improving the quality of writing in an organization.

¶ The sixth responsibility of management is *to establish and insist on promptness* in report writing.

By their nature, some reports must be written to a deadline. Proposals and monthly progress reports are outstanding examples of this kind of writing. (Of course some laboratories call for more frequent regular reports, but since the value of such reports is debatable, the practice will not be discussed here.) By hook or by crook, writers usually manage to meet these deadlines. Because the time a man has for writing a proposal is usually too short, the quality of writing often suffers. Management has the responsibility in such cases of trying to make more time or writing by adequate staffing and efficient supervision of the proposal. In cases of reports for which the deadline is less severe, the problem of management is quite different.

A research report does very little good while it is in process of being written. By the same token, it does very little good while it lies in a desk drawer waiting for editing or final typing. But this latter problem has already been discussed. Therefore writers must turn in their reports in reasonable time. The decisive factor in determining whether a report will be done in good time is the pattern of supervision established in a group.

From the time of the first conference between writer and supervisor before the writing of the first draft, a supervisor must make clear just when he expects a report to be delivered, and he must not sit back and hopefully wait for it.

While some supervisors make the mistake of nagging to the point where a writer dreads the sound of footsteps outside his room, others are so timid that they cannot bring themselves to the point where they dare ask how a report is coming along. Somewhere between these extremes lies good management.

One way to approach the problem is to ask for the report in parts. As a man finishes the first draft of a section, he sends it along to his supervisor. This plan can work well if it is handled properly. It depends first on whether a good outline has been developed by the man, with or without the help of his supervisor. With a copy of the outline next to him, a supervisor can see where a section fits into the over-all scheme and he can read intelligently without sending back useless queries, asking whether this matter or that has been dealt with prior to the section he is reading, and whether the writer plans to cover still another topic. It also depends on how quickly the supervisor is able to send the section back to the writer, either with a compliment or with other types of comments. If the supervisor sees that major reconstruction is necessary, it is far better for him to get onto the problem with just a sample of the writing than to wait for the entire report before getting out his shears and sutures. Hopefully, any corrections made on the early sections of the paper will help the writer avoid making the same kinds of errors later on in their report.

Some writers do not like to submit a report piecemeal. Whether this feeling reflects poor planning before writing or general fear of writing does not matter. The supervisor can still accomplish his purpose of seeing that a report is coming along on schedule by dropping in on the writer once in a while to ask how things are going.

When a report is being written by several writers, it is especially important to see that a schedule is being kept. A delay in one small part of an effort delays an entire project. A false start in one part of a report may require such extensive rewriting late in the game that the entire report will be held up.

In saying that a tight schedule never hurt a report, we do not condone a cruelly tight, well-nigh impossible one. If management will see to it that a man's other duties do not interfere with his writing, he will do as fine a job under the pressure of time as he will with all the time in the world. In fact, the longer some writing is delayed, the worse it gets, because the information begins to fade rapidly in vividness and intensity the further a man gets from his work.

MANAGING A PROPOSAL

The proposal is a document designed to sell the services of the bidder. If it is to do this job effectively, it must convey to the evaluator the feeling that the bidder is competent to do the job, has thought it through carefully, has personnel who have solved similar problems, has equipment and plant for doing the job well, and has a history of meeting production and delivery schedules.

For a bidder pre-eminent in a field the task is not difficult. For another bidder the job is sometimes extremely hard. He cannot fill pages with lists of accomplishments. The experience of his company and its personnel may not match that of other bidders. Yet if the decision to bid has been soundly made, there is a course open to him in formulating his proposal.

Honesty reflected in straightforward discussion of the technical problems involved will do the most to assure favorable consideration by the evaluator. Even though the bidder may not have direct experience in the field he is trying to enter, he has a background in tangential areas. He has devoted company funds to study the central problems of the technical area. He has personnel who have spent much time on these problems. His related experience can be presented with promise of consideration.

The writing style of the proposal should convey this attitude of the bidder toward the subject. Glittering claims are rapidly revealed as so much wind. Extraneous discussion irritates the evaluator and does not fool him.

Proposal effort

Preparation of a major proposal is an expensive and time-consuming task, involving engineers, salesmen, managers, artists, typists, and printers. It is a time effort. Good organization from start to finish saves wear and tear on personnel. It results in effective proposals.

And the secret ingredient in the effort is time.

Given unlimited time, most engineering companies can produce a sound and effectively presented proposal. But not only is there never unlimited time, there is never enough time. One way to make time is to initiate and sustain engineering studies long before an invitation to bid ever appears—as the readers of this book know better than its authors. But *time can be made if a well-organized proposal effort begins with arrival of the invitation to bid*.

The first essential in a sound proposal effort is an early decision on whether or not to bid. This requires a technical sales force that is on its toes so that no invitation to bid takes an engineering staff completely unaware. It also demands a well-thought-out policy concerning the company's areas of concentration. Bidding on every invitation is a sure path to failure. Bidding only on certainties means a rapid shrinkage of a company's scope. Somewhere between these extremes lies the proper bidding policy for any company. This matter of policy is best left to the top engineering and management echelons.

Surely an invitation allowed to gather any dust had better not be bid. Little enough time is allowed from issuance of an invitation to receipt date for a proposal. Any delay in decision to bid places undue hardships on a proposal team.

In a typical situation, the team has about three weeks from the time the bid request is received to delivery day. Considering the amount of study and discussion needed to arrive at a design plan, there are precious few days left for writing.

The committee charged with deciding on bid requests must be ready to meet daily if the needs of the proposal team are to be considered. Along with the decision on whether to bid, this committee should have the responsibility of selecting a proposal chief. With these two decisions made, the committee can go on to its other duties until the next bid request arrives.

The proposal team

Proposal teams should be as large as a company can afford. If the members of a team are going to be able to think through a technical problem and come up with a written document supporting its thinking in a matter of days, the task must be broken down into many small parts.

To act effectively, however, the team must have a chief capable of directing every part of the effort. Naturally he must be a person who has the broad technical knowledge that will enable him to lead and evaluate the work in progress. While he does not have to know as much about all the details of the problem as the men under him, he must be able to understand and co-ordinate the thinking. He must also be able to run the show in such a way that the men on the team will put forth their best efforts and be willing to serve under him again when he is selected to lead them in a new proposal effort. Finally, he must himself be a good writer and editor.

The chief's first job is to read carefully through the bid request and specifications and select senior members of the staff to work with him on an outline and to recommend groups of writers to assist in the various parts of the proposal. He must see that all the members of the writing team he gathers are made familiar with the broad outline of the job and that they receive the detailed sections of the specifications that they will be concerned with.

The best procedure is to call the entire team together for a preliminary meeting, at which the chief presents a tentative outline of the proposal, with copies for each writer. This outline is much more than a listing of the topics to be covered in the proposal. What is required is at least a sentence or two of discussion of each item to guide the writers in their thinking and writing. Outlines that merely list items to be covered have caused more wasted effort than anyone unfamiliar with proposal writing can imagine. Unnecessary duplication of effort, omission of vital discussion, overwriting, and underwriting result from inadequate outlines.   .

Properly organized and presented, the outline is an effective first step.

The chief assigns each *section* within the outline to a senior staff member for supervision, and he then assigns each *subsection* to one man for study and writing. In the discussion at the meeting, each entry in the outline is thoroughly explained by the chief and by all others who wish to speak. No meeting should be concluded before all doubt has been removed as to what is going to be done.

The chief gives two other items to each member of the team: the time schedule for the proposal effort and the preliminary draft of the proposal abstract. The reason for the schedule is obvious. Everyone on the team must know when his portion of the material must be in the hands of his section chief. Every section chief must know when he must hand his material in to the proposal chief. The dates assigned must be realistic, taking into account the time required for study and writing and the time required for editing, typing, and printing. In actual practice, the latter considerations set the limits for the schedule, and the writers must make do with what time is left.

Giving a preliminary draft of a proposal abstract to the members of the writing team has several advantages. First of all, every member of the team sees quickly how the entire proposal is intended to shape up. He sees where his contribution fits into the over-all picture. He acquaints himself with the tone and emphasis of the proposal, so he can develop his section in appropriate style. Finally, he knows that the project is really under way when he has in hand one of the most important parts of the document. He is ready to go to work.

If the proposal chief is to function effectively, he should have an assistant who will be helpful in many ways throughout the proposal effort. This person need not have the broad capabilities of the proposal chief. He can be put in charge of the management section of the proposal, which is often collected from existing documents. If any new management material must be written, he can undertake that. His most important task will be to take administrative detail off the desk of the proposal chief. He must see whether work is proceeding on schedule and keep his chief informed of progress and bottlenecks so that additional help can be obtained or some writing reassigned.

There is no question about it: the jobs of proposal chief and assistant are full time.

As first drafts begin to come in, the proposal chief functions as sole editor. The section chiefs have already read the material for technical accuracy, and the proposal need not go through any more hands. The chief keeps his records of material received and, as soon as he finishes with a section of the proposal, sends it off for final typing preparatory to printing.

Others are preparing the financial and scheduling data that usually accompany a proposal, but the proposal chief stays abreast of this work too, in order to make certain that it will be ready in time.

If a laboratory develops a pattern of perpetual overtime and last-ditch measures in proposal preparation, one of its biggest problems will be finding people of outstanding caliber to act as proposal chiefs. If the laboratory finds a way of managing its proposal efforts with a minimum of blood, sweat, and tears, the quality of writing and thinking that go into proposals will improve, and so will the company's batting average. The procedures described for staffing, supervising, and scheduling proposals may help.

SOLVING THE READING PROBLEM

In Chapter 1, a *long abstract* was described that makes it possible to limit circulation of full reports of development and research. Those scientists and engineers who must see full reports would receive copies, but those who do not normally want to plow through a lengthy report would be informed of current findings without the kind and amount of detail they do not care to read. Some further discussion of the printing and circulation of reports is needed in our consideration of the management of technical writing. It is hoped that the discussion will lead to mitigation of the reading problems of engineers and scientists.

The advent of high-speed mimeographing made possible the inexpensive printing and distribution of as many copies of reports as there are men in a given organization. Introduction of this service was welcome, but as time went by and the number of men in the typical laboratory increased, distribution lists for reports grew and grew. The end is nowhere in sight. How can a report writer be sure that a man left off a distribution list really does not want to see a report? Isn't it possible that omission of certain names from the list may be tolerable 99 per cent of the time but constitutes a serious error in the remaining one? Better circulate to everybody and take no chances—mimeographing costs are low.

This reasoning was satisfactory until the point was reached in many organizations at which the amount of company mail deposited on desks each day was so large that *men ran out of reading time,* even though mimeograph facilities were cheerfully working away well within capacity. Reading techniques adopted widely by busy men rely on a quick look at titles, a hurried skimming of abstracts, a glance at conclusions. This is unsatisfactory. A new solution is clearly needed.

The key to the solution of the glut of unread reports that is threatening to overwhelm our scientists and engineers lies in the new photocopying machines that are now commonplace in industry. Full reports of studies should be typewritten in three copies: one for the writer, one for his supervisor, and one for files. A long abstract should be circulated to everyone on the technical staff of an organization. If a reader decides that he would like to know more about the study reported, all he has to do is request a photocopy from files. Presto.

The effort involved in asking for the copy will deter the mere report collector. No more will he accumulate and accumulate for the sake of accumulation. File drawers in individual offices will stand empty. Overburdened company mail deliveries will be relieved. Even costs may go down.

The abstracts that will be widely circulated can be written and distributed quickly because they need not wait for completion of the full report. Dissemination of valuable information will probably be improved.

THE WRITING SEMINAR

More and more companies look upon the continuing education of their employees as a responsibility of management, expressed in most cases as some sort of tuition-payment plan to encourage men to attend classes at a nearby university. In others, and the number seems to be growing, courses are conducted in the plant or laboratory. The advantages of this latter scheme are many, the most important being the possibility of courses organized to meet specific needs.

A writing seminar is one such course.

Rather than have scientists and engineers attend a university course in writing that attracts a wide variety of students with a corresponding variety of backgrounds and needs, it is sensible for a laboratory to organize its own writing seminar. The course content can be limited to the specific needs of the participants, and the instructor can dovetail with company requirements.

A few guideposts are essential for laboratories contemplating such a program.

- The seminar must be limited in size. There is little value in listening to lectures on writing, and lectures are inevitable in large groups.
- The textbook must be the articles, memos, and proposals written by members of the seminar. Nothing is so revealing of the inadequacies and strengths of a man's writing as listening to his associates criticize it.
- Every member of the seminar must be actively engaged in writing, so that his work can be criticized.
- Homework assignments must not be made unless a writer wishes to rewrite a report, or part of a report, as a result of criticism he has received.
- If possible, the seminar should be held weekly for a period of two hours or more. It should meet during working hours because it concerns work. Since a group of eight men can usually be helped by an eight-week course, not many hours will be needed to complete training.
- The instructor must be a person skilled in writing and teaching. He should be thoroughly familiar with the kinds of writing done in the laboratory. In the ideal case this person should be a member of the laboratory staff. As a practical matter, however, the combination of abilities and knowledge is not always available, and there is the risk of offending some members of a seminar by criticism that is too harsh. When an instructor cannot be found in a laboratory staff, an instructor can be recruited from a nearby campus. If he is given proper guidance, he can quickly learn the laboratory writing requirements and serve a useful function.
- No grades must be given. The objective of the seminar is to help each writer advance as far as he can, regardless of how proficient he is at the start.

With such a program, the level of writing and editing in a laboratory can improve markedly. Inevitably, the quality of management of technical writing will also improve.

A HANDBOOK OF STYLE AND USAGE

GETTING STARTED

An author once said that there was no experience more terrifying in life than the terror of the blank page. Most persons would not use the term terror, but anyone who has even written the simplest article will confess that getting a paper started is difficult: "If I could only get the first page written . . ."; "Once I get started I'm all right . . ."; "Getting started is the problem. . . ." Getting started *is* the problem. Once started, a writer who has his material in hand will find it easy to write the article or report.

Although the terror of the blank page can never, perhaps, be overcome, it can be alleviated

- by thinking of your audience
- by limiting the focus
- by writing a theme statement
- by writing an outline

THE AUDIENCE

Nothing exasperates a reader more than to be told something he doesn't need to know or not to be told something he needs to know in order to understand a report. The potential reader should be kept in mind when the writer lays out his paper. "Who is going to read this paper?" "What is his level of competency in the subject?" are questions the writer should ask himself before he begins a paper. Different audiences require different approaches. A paper written for engineers on New York buildings differs from one written for contractors, for prospective buyers, for architects, or for the idly interested. A field trip in the woods would have to be described differently

if the report were addressed to an ichthyologist, to a lepidop-
terist, to a botanist, or an archaeologist.

WRITE THE PAPER FOR YOUR AUDIENCE

"How much of this subject can I presume the reader
knows?" "What does he need to know to make the subject
perfectly intelligible to him?" The answers to such questions
indicate the level on which the paper should be written.

Limiting the audience immediately limits the area. Jotting
down on a sheet of paper the possible topics to be covered
will help determine the area of the paper. Some of the topics,
when examined for possible inclusion, will seem either too
elementary or too complicated; others too far removed from
the central concern of the paper. These topics should be
crossed out, and those that remain should be grouped by giving
those that fit together similar numbers (such as 1A, 1B, 2A,
. . .). These preliminary groupings will be helpful in formulating both the topic statement and the outline.

Here is a paragraph from a book, *How Animals Move* by
James Gray, which was directed to a juvenile audience. Notice
how the audience determined the level of fruitful discussion
and how circumscribed the area becomes when the audience's
competency is considered. Do you think anyone with a good
knowledge of anatomy would be satisfied with the sentence,
"Each fibre is supplied by a nerve coming from the brain . . ."?
Yet had the author intensified his focus on this area, he would
probably have lost his audience.

> The movement of all the larger animals depends on
> muscle fibres, for these structures are by far the finest
> and most powerful of all Nature's engines. The muscles
> of our bodies are made up of a number of these fibres
> bound together, and each fibre, though sometimes several
> inches long, is not more than a very small fraction of an
> inch in diameter. So long as its fibres are at rest, the
> muscle is soft, flabby, and can easily be stretched. Each
> fibre is supplied by a nerve coming from the brain, and
> when impulses pass down this nerve, the whole character
> of the muscle changes. It shortens in length and strongly
> resists any attempt to prevent this happening. It changes,
> in effect, from a piece of rubber to a stretched steel spring.
> As soon as the stimulus from the nerve ceases, the muscle
> becomes limp again; but it does not stretch itself out to
> its original length; to regain length the muscle must either

be pulled out by the contraction of another muscle or extended by some other force not its own.*

The preceding selection derives from six lectures the author presented to a juvenile audience at the Royal Institution, Cambridge, England. Since the audience was juvenile, the focus is broad and no area is investigated to its full potential; however, the inclusion of terms such as "stimulus" and "impulse" indicates that the author was not in any manner talking down to his audience.

Here is another selection, this one from *Natural History,* a magazine designed for general distribution by the American Museum of Natural History. Notice that the author has refrained from presuming subject competency by the audience; he has presumed only that the audience is familiar with the animal:

> As the giraffe paces, its neck and head move back and forth twice during each stride. They reach their farthest point forward just before the right legs touch the ground and again just before the left pair is set down. Thus, the forward impulse of the neck shifts the animal's center of gravity to the front and aids the advance of the body. As each pair of hoofs touches the ground, the head is drawn back again to decrease the giraffe's forward momentum and to prepare for the next frontal thrust of the neck.

Notice how much more general the discussion is here than in the previous passage on animal movement. In this passage the focus is very broad, for the audience cannot be presumed to have the level of competency assumed for the audience of the first selection. Remember: no paper should be started until the audience has been determined; only then can the writer know at what level and with what kind of focus the paper should begin.

WRITING A THEME STATEMENT

The first task in writing a paper is the formulation of a theme statement. Ask yourself, "What is the main idea that I wish the reader to understand?" The answer to this question *should be written out* as specifically as possible. Vagueness is the death of good ideas. To use a simple example of a topic statement, let us suppose we were asked to write a paper on

---

* By permission. From *How Animals Move,* James Gray, published by Cambridge University Press.

how to hit a tennis ball. The audience is presumed to have seen a tennis match, but never to have played the game. What does this reader need to know about hitting a tennis ball? He needs to know that there are four basic ways to hit the ball:

> Four basic strokes are required to hit a tennis ball properly: forehand, executed at right angles to the ball, with the left foot forward (use illustration from baseball?); backhand, executed at right angle to the ball, with right foot forward; the volley, executed either forehand or backhand, with the arc of the stroke shortened; and serve, in which the left foot is placed on the serving line and the ball tossed higher than the head and slightly in front of the server. (Note: what about left handed players?)

Three matters should be noted in this sample topic statement. First, the topic has been divided into four equal parts, facilitating outlining; secondly, the parts have been stated specifically enough so that the writer is certain of what he wants to say; and thirdly, ideas that occurred while the theme statement was being written have been jotted down so they will not be forgotten and lost.

If possible—and it is often possible if the jotting down and grouping process recommended above has been followed—the theme statement should be formulated in such a manner that the logical divisions of the subject are apparent. Often a theme statement can be formulated if the writer uses the debating term: "Resolved that. . . ."

FORMULATING AN OUTLINE

After the theme statement has been written as concisely and as clearly as possible, an outline should be developed from it. Outlines vary according to the nature of the subject, but all are basically of two types—the schematic and linear. In the schematic outline, lines indicate relationships. In the linear outline, the main headings are usually given Roman numbers (I, II, III, IV), with the subheadings in capital letters (A, B, C, D); details of the subheadings are in arabic numerals. Schematic outlines are easier to construct, contain fewer details, and are easier to read than linear outlines. For material which is familiar and for which the writer has adequate detail and structure, the schematic outline is very effective. For less familiar material, which requires considerable thought before structure can be imposed on it (more correctly, derived from it), the linear outline, with its more fully developed treatment, is much superior. The choice of outline form depends on the material and the preference of the writer. The use of a preliminary schematic outline, with a fully developed linear outline derived from it, is very effective for organizing complex subjects.

Schematic outlines

A schematic outline enables the writer to divide his material into its major and minor classes or divisions and to structure these relationships easily and rapidly. To construct a schematic outline for the topic used earler, "How to Hit a Tennis Ball," the term *strokes* will allow us to divide the concept into logically equal parts; the term *position for hitting* will give the subheadings.

How to Hit a Tennis Ball

Strokes ..... Forehand	Backhand	Volley	Serve
Position			
for hitting...body racket	body racket	body racket	body racket

The main use of schematic outlines is for subjects that divide logically into classes. One of their faults is that the writer is not required to formulate the division of the subject into either phrases or sentences; he is required only to indicate the relationships.

Linear outlines

Linear outlines are of two types—the phrase outline, in which the statements in the outline do not have a subject and a verb, and the sentence outline. Either is very effective, but the sentence outline, requiring more fully developed statements, is a great aid when the main heads are converted into topic sentences. Since most of a writer's time is not spent on writing, but on thinking through a subject, an outline is a sort of shorthand for getting his thoughts on paper. Thinking when the outline is being constructed saves time, since relationships are easier to change when they are in outline form. Time spent formulating an outline will be time saved in writing the paper.

One of the most important points to remember in writing

an outline is that the subject must be divided into equally important sections. Improper division will result in improper emphasis. The main divisions of the subject should be clear if the topic statement has been formulated precisely. Here is the beginning of a linear outline for "How to Hit a Tennis Ball." It is in sentences.

I The forehand is the basic stroke in tennis.
   A The body should be at right angles to the net, with the feet about 18 inches apart, left foot forward.
   B The racket should be held in the right hand, with the arm extended well behind the right foot.
II Use a backhand stroke for balls that cannot be reached with the forehand.
   A The body. . . .
   B The racket. . . .

During the outline stage of writing a paper, the writer should examine the divisions of the work and the logical relationship of the parts. It is at this stage that structural corrections should be made. It is much more difficult (and costly) to do so after the paper has already been written. In the example above, some writers would argue that logically the stroke that puts the ball in play (serve) should be discussed first. Such a change could readily be made in the outline. *Outlines save time and effort.*

Elsewhere in this text you were given detailed instructions on the organization and execution of the paper or report. If the reader has not yet read the section on the introduction, body, and conclusion (pp. 50-56), he should do so before he begins to write the paper. He should also keep in mind the three cardinal virtues of all writing: unity, emphasis, and coherence. Unity means that a paper should be about one subject —the concept stated in the theme statement; other material should be saved for another paper or report. Emphasis means that the most important concepts occupy places of greatest importance in the paper: the introduction, the topic sentences of the body, and the conclusion; the subject and verb of each main clause. The paper must also be coherent. The place to impose coherence on the paper is the outline. It should be examined to determine if all the paragraphs derive from each other. If the writer could just as easily put the third paragraph in place of the second, there is something wrong with the organization. Coherence also means that each paragraph is linked to the preceding one by some transitional device; that each sentence is logically related to every other sentence, and that the elements in each sentence are coherent—that they hang together.

In the following pages the reader will find chapters on the paragraph, the sentence, the word, and the verb. He should refer to them as well as to the other sections on grammatical and stylistic usage as aids.

STRONG, WELL-DEVELOPED PARAGRAPHS

A paragraph is one of the major building blocks of a paper. If the individual paragraphs are forceful, specific, and adequately detailed, the chances are that the paper will be good. All good paragraphs have three elements:

- A transitional word, phrase, or clause that links a paragraph to one that precedes it.
- A topic sentence.
- A group of sentences that document, support, or prove the statement in the topic sentence.

THE TOPIC SENTENCE

The topic sentence is a statement that introduces the subject of the paragraph. It states a general concept that constitutes a division of the paper's theme. Normally topic sentences are developed from one of the main headings of the outline, and the details necessary to support the topic sentence come from the minor headings—the subheadings.

Write the topic sentence with these details in mind. A topic sentence introduces a *topic.* Make it concrete and complex enough so that details *must* be included to prove the validity of the topic statement. At the end of the paragraph, the reader should be as convinced as the writer of the truth of the topic sentence.

The key to good paragraphs is the topic sentence.

- Write the topic sentence from the outline.
- Write the topic sentence with the supporting details in mind.

- Write it as concretely as possible. A paragraph is a development of the topic sentence. *Nothing* develops from nothing.

The following two examples should be read carefully. The first represents the most common fault of the topic sentence—the failure to write strong, concretely detailed statements that introduce the reader to the main idea of the paragraph. Suggested corrections are in the second sentence.

> The study of eclipses has contributed *a great deal* to the advance of science. In 1919 an expedition to Brazil took photographs of the sun during an eclipse *to check on one* of the predictions made by Albert Einstein in connection *with one* of his famous theories—the theory of relativity.
> In 1919 a Brazilian expedition that photographed an eclipse provided proof of Einstein's hypothesis that light bends when it passes a large mass such as the sun.

In the first example, the reader is required to read two sentences without learning anything specific. Such phrases as *a great deal, to check on one,* and *with one* have no meaning. The actual subject of the paragraph, the proof that light bends, hasn't even been mentioned. There is no excuse for writing such vague sentences. Topic sentences that do not introduce the reader to the topic waste the reader's time. The second example provides the reader with all the information he needs; the writer can now marshal his details to prove the statement of the topic sentence. Here is another double example:

> From the total wealth of impressions received from nature, Copernicus, Kepler, and Galileo fastened upon *some only* as being suitable for scientific formulation. All three scientists used these as the basis for their scientific formulations.
> From the total wealth of impressions received from nature, Copernicus, Kepler, and Galileo fastened upon those that possessed quantitative aspects as being suitable for scientific formulation.

The first example does not mention the only significant term, "quantitative." Without that term, the reader cannot possibly understand what the paragraph will deal with.

Make certain that all your topic sentences *assert* something concrete about something specific.

Significant details

Once the topic sentence has been adequately formulated, the reader must be given the facts upon which the hypothesis depends for its validity and be convinced that the concept in the topic sentence is true. Therefore the writer must take the reader step by step through the process he himself used to arrive at the hypothesis. Every sentence in the paragraph should be logically necessary and related to both the preceding and following sentences; every sentence in the paragraph should be where it is because logical necessity dictates that it occur where it does.

Refer to the outline for the details and the order in which they should occur. If, after using the outline and writing the details as concretely as possible, the paragraph does not seem to hang together or to develop consistently, use the following suggestions as a guide in reorganizing the structure.

- Try to impose a pattern on the material—temporal or numerical sequences are often possible. Frequently an easily recognizable pattern will help, such as the points of the compass or from left to right, front to back, up or down, inside to outside, and so forth.
- Use a noun in the preceding sentence as the subject of the new sentence: The man swung at the *ball* with all his might. The *ball* traveled only two feet.
- Use appropriate conjunctive adverbs to connect sentences and independent clauses: *moreover, however, nevertheless, consequently,* and the like.

Read the following as an example of a paragraph that uses all the devices suggested above to achieve a unified, well-documented paragraph that convinces the reader that the statement made in the topic sentence is valid. It moves well because the author has picked a point on the leg and moved from that point to another in a logical manner:

> The leg of a mule shows three main differences when contrasted with a bear's leg. *First,* the five toes in each limb of a bear help to provide a broad-bearing surface, but the limb of a mule consists of only one toe, and this ends in a small and rigid hoof. *Secondly,* neither the heel ("hock") nor the wrist ("knee") of a mule ever rests on

the ground; these joints are carried well away from the ground because of the length of the bones uniting the "finger" to the rest of the limb. *Thirdly,* the limbs of the mule are longer than those of a bear, and the muscles that work them are concentrated at their upper ends, the lower joints being operated by long tendons. Unlike a bear or a man, a mule supports itself on the top of the middle toe of each foot; all the other digits have disappeared.

All the details in the preceding paragraph contribute to meaning. They are concrete and specific; moreover, it is quite apparent that the writer knew the details *before* he wrote the topic sentence—in other words, he was working from a well-developed outline.

TRANSITIONS BETWEEN PARAGRAPHS

Each paragraph must be joined to the preceding one in such a manner that the reader understands the relationship between them. This transitional element is normally a word, phrase, or clause at the beginning of the new paragraphs; it is frequently attached to the topic sentence. There are two means for accomplishing the transition:

- using a conjunctive adverb or adverbial expression
- repeating the main idea or an important term from the preceding paragraph

Transitional adverbs

Frequently the material in two paragraphs is so closely related that one of the following transitional adverbs may serve as an effective transition: *moreover, consequently, however, additionally, in addition to, first, second, subsequently, furthermore,* etc. Any of the foregoing adverbs can be used in front of the topic sentence. All are normally set off by a comma.

> *Furthermore,* the horse differs from a bear . . .
> *Moreover,* the horse differs from a bear . . .
> *In addition to* the difference in leg structure, a horse differs from a bear . . .

Summarizing a main idea or repeating a key word

Either a phrase or a clause is used to make a transition that is not achieved solely by the use of a conjunctive adverb. Both the clause and phrase usually precede the first sentence of the paragraph and are set off from it by a comma. For example, in the paragraph on the leg of a mule (see p. 94), a transition might be made by using the important term *differ*.

> While the leg of a mule *differs* from a bear's, both animals are similar in . . .

Using a phrase from the last sentence of the paragraph will also make a smooth transition.

> Although *all the other digits* of the mule's foot *have disappeared*, there are still some similarities between the manner in which the mule and the bear walk. . . .

PARAGRAPH FORM

Most material developed in paragraphs is dealt with in one of these five expository forms: definition, classification, comparison and contrast, analysis, and illustration. An understanding of these forms and the kinds of material best developed in each will simplify the order and arrangement of details and will facilitate the imposition of a logical pattern.

Definition

Whenever a writer undertakes to tell a reader what a word, or a group of words comprising a unit, means, he is using definition. Writers are constantly defining their terms, so it is useful to be certain that when a term is defined, the reader understands the limitations placed on it.

A good definition is formulated by locating the general class to which the term belongs and then distinguishing that term from all similar ones in the class. On the simplest level, a definition results from the question, "What is it?"

> What is a transistor? A transistor is . . .

Complications arise, however, when the class to which the term is related is not known to the audience. If, in the example above, a transistor had been related to the class "electron tube," with which the audience was unfamiliar, it would be impossible to define "transistor" without first defining "electron tube" by relating it to some class with which the audience is familiar. An effective definition, then, not only relates the term to the class to which it belongs, but also recognizes that effective definition can only occur when the audience level is understood.

Here is a paragraph by Thomas Huxley that answers the question, "What is chalk?"

> We know that if we "burn" chalk the result is quicklime. Chalk, in fact, is a compound of carbonic acid gas and lime, and when you make it very hot the carbonic acid is distilled off and the lime is left. If, on the other hand, you powder a little chalk and drop it into a good deal of strong vinegar, there would be a great bubbling and fizzing, and finally, a clear liquid, in which no sign of chalk would appear. Here you see the carbonic acid in the bubbles; the lime, dissolved in vinegar, vanishes. There are a great many other ways of showing that chalk is essentially nothing but carbonic acid and quicklime. Chemists enunciate the results of all the experiments which prove this, by stating that chalk is almost wholly composed of carbonate of lime.

Classification

Classification is a form of paragraph structure used very frequently in technical writing. As its name implies, classification is the breaking down of items into the classes to which they belong. Definition isolates the term being defined from all terms that seem like it; classification relates the term or item to those items that are similar. Definition seeks differences; classification, similarities. The books in libraries are arranged by classification: all books on history are in one section, and these are then broken down into subgroups—ancient history, medieval history, modern history, and so forth. Similarly, all plants and animals are divided into groups and then subdivided into lesser groups.

Comparison and contrast

Comparison is used to indicate the similarities between two essentially different items; contrast, the differences between two items essentially similar. Normally comparison is used when the writer wants to describe something relatively unfamiliar to the reader. He describes the unknown in terms of the known: to describe a little-known marsupial such as the wombat, the writer might wish to compare it with the better-known kangaroo; he would use contrast to show the differences between these two animals having a common characteristic. When using comparison and contrast, be certain that the things compared have common characteristics. Also remember your reader's level—don't compare something he doesn't know to something else he doesn't know.

Analysis

Analysis deals with the relationship between cause and effect: given this set of causes, what results will follow? given this result, what could the causes have been? Causal analysis is normally concerned with the *immediate connection;* all connections ultimately end in Genesis. All of us use causal analysis: when hungry, we eat, for we know that hunger is caused by lack of food; eat and the hunger disappears. However, when we are dealing with analysis in a paragraph or a paper, we must be certain that the relationships are dealt with systematically. The reader must be led logically, step by step; there should be no backtracking to fill in some detail omitted earlier but later found necessary.

Illustration

Like classification, illustration is a means of thinking in classes. However, illustration differs from classification because it takes a typical item from the class and, by describing that item, describes the class. Thus, if the writer wants to describe the class "butterfly," he may take any particular butterfly, such as a tiger swallowtail, and describe it, omitting all those peculiarities of the tiger swallowtail that distinguish it from other butterflies in its class. The reader then understands the general characteristic of the class rather than the particular characteristics of the tiger swallowtail. Illustration also differs from classification in that its techniques are readily adaptable to expository narration, a form in which narrative techniques of character, temporal sequential order, and dramatic conflict can be used to explain phenomena. Thus if the writer wants to explain the characteristic style of a group of French Impressionists, he might take a reader to a hypothetical lecture on Monet, and, using a lecturer and audience, describe Monet's painting in a manner that enables the audience to understand French Impressionism and, incidentally, Monet.

It should be quite obvious that we rarely find any of the major expository forms in a pure state; comparison and contrast are sometimes utilized in classification, and so forth; however, there will always be one form that is the major form. The writer should keep the major pattern in mind and give the paragraph or paragraphs as precise a pattern as the major form will permit.

THE SENTENCE

A sentence is a structurally complete unit of meaning, composed of a subject and a predicate. The subject is the initiator of the action, and the predicate—the verb and whatever is needed to complete the verbal action—indicates what the action is.

SUBJECT	PREDICATE
The *tube*	*exploded.*
The *fire*	*burned the house down in twenty minutes.*

A group of words can only be considered a sentence if it has a subject and a predicate. A phrase such as "moving slowly through the laboratory" is not a sentence because it does not have a verb or a subject—the term "moving" is a verbal, but it does not state that any action was initiated by a subject. A sentence must also be a complete thought. The phrase "which he had left behind" does not constitute a sentence, although it contains a subject *(he)*, and a predicate *(had left behind)*, because it does not contain a structurally complete unit of meaning. It depends on some other structurally complete sentence element to give it meaning:

The *stick* which he had left behind *disappeared.*

A sentence, then, is composed of two elements—a subject and a predicate that express a complete thought. If the verbal action is transmitted to any element in the sentence, that sentence element is the *object* of the sentence:

The battery started the *engine.*

If the sentence has an indirect object—one that names the receiver of the object of the sentence—it is placed between the verb and the object. This relationship of subject, verb, indirect object, and object is the fixed order of an English sentence.

The relationship of any of these fixed sentence elements cannot be changed without either changing the meaning or the emphasis, or causing meaningless sentences:

The man hit the ball to him.
The ball hit to him the man.

Since meaning is a grammatical function, discussion of the sentence as a meaningful entity begins with the means for finding the basic sentence elements. To find the subject of the sentence, take the main verb and ask *who* or *what* plus the verb: Who *hit?* The man hit. . . . What *hit?* The ball hit. . . . To find the object, take the verb and ask *what* or *whom*: Hit *what?* The ball. Hit *whom?* The man. The indirect object is the person or thing receiving the object.

The ability to identify the main sentence elements is necessary because all other sentence elements depend on them for meaning.

MODIFICATION

Normally the main sentence elements must be defined or limited to some degree. This function is performed by adjectives and adverbs. An adjective is a word or group of words used to modify, define, or limit a noun; an adverb modifies a verb, an adjective, or another adverb. Adjectives and adverbs may be either one-word modifiers or multiple-word modifiers —phrases or clauses.

Single-word modifiers

Single-word adjectival modifiers precede the word they modify: *blue* hat; *extruded* aluminum; *progressive, product-oriented* research laboratory.

Single-word adverbial modifiers can normally be moved freely in the sentence (this also holds for complex adverbial modifiers):

He searched the laboratory *carefully*.
He *carefully* searched the laboratory.

Complex modification

The Phrase—A phrase is a group of words without a subject or predicate; it performs the same function in the same manner as single-word modifiers. Thus, in the sentence

> The chief engineer of the company worked in the laboratory for an hour.

The phrase "of the company" modifies engineer in the same manner as "chief"—it is a prepositional phrase used as an adjective. The prepositional phrases "in the laboratory" and "for an hour" modify the verb. They are therefore adverbs.

A prepositional phrase has two elements, the preposition and the object of the preposition: *in* the *house, from* the *lab, to* the *company, from* a highly classified company *document.*

Verbal phrases also modify other sentence elements. A verbal is a verb form that functions as a noun or adjective. The principal verbal phrase is the participial, which functions as an adjective; it can be identified by remembering that the present participle (ending in *-ing*) not preceded by some form of the verb "to be" is probably a verbal; if it is preceded by some form of the verb "to be" it is usually a progressive tense.

verbal phrase:	*Mixing the solution* as he consulted his chart, the chemist made a dreadful error. (Adjective phrase modifying "chemist.")
progressive verb:	As he *was mixing* the solution and *(was) consulting* his chart, the chemist made a dreaful error.

The past participle (normally ending in *-ed*, except for irregular verbs: pour*ed*, but *run, gone*) is a verbal if it is *not* preceded by some form of the verb "to have."

verbal phrase:	*Watched* by an attentive audience, the surgeon entered the operating room. (Adjective phrase modifying "surgeon.")
verb:	The attentive audience *had watched* and *(had) waited* for an hour before the surgeon entered the room.

The infinitive is the simple form of the verb, normally preceded by "to": *to run, to hope.* Infinitives are used both as adjectives and adverbs in addition to being used as nouns.

adjective:	He asked the consulting engineer *to lead* the discussion.
adverb:	He poured hastily *to fill* the flask.

The Clause—Clauses are also used to modify. When a group of words contains a subject and a predicate but does not constitute a complete unit of meaning, it is a dependent clause, depending on some other element for its meaning. Dependent clauses may function both as adjectives and adverbs.

> adjective: The most significant statement *that the lecturer made* was that two plus two are four.
>
> adverb: He wanted to know *where the package should be placed.*

SENTENCE TYPES

Writers use—usually without knowing it—four different sentence types: simple, containing no clause; complex, containing one or more *dependent* clauses; compound, containing two or more *independent* clauses; and compound-complex, two or more independent clauses containing one or more dependent clauses; this last type is, however, no different from the complex if one treats the independent clause as a unit. It is not discussed below.

¶ The *simple sentence* rarely presents a problem to a writer, except when he uses it exclusively, producing primer sentences:

> The company hired the engineer. The engineer worked very hard. The company liked the engineer's work. It gave him a better job. The engineer worked harder.

Such sentences should be changed into more complex forms to avoid unnecessary repetition and to subordinate sentence elements, as below.

¶ The *complex sentence* subordinates one sentence element to another and eliminates primer style:

> adjective: The company hired the engineer, *who worked very hard.*
>
> adjective: The company, *since it liked the engineer's work,* gave him a better job.
>
> adverb: He worked *whenever the company needed him.*

Not all subordinate clauses are modifiers—some are main sentence elements. Here is a subordinate clause used as subject:

That the company appreciated his work was evident.

¶ *Compound sentences* have two or more independent clauses of grammatically equal value, one of which is normally a consequence, explanation, or continuation of an action begun in a prior clause. Each of the independent—coordinate—clauses must be separated by either a semi-colon or a comma and by one of the coordinating conjunctions—*and, but, for, or, so, yet, nor.*

The company liked his work, *so* it gave him a better job.
The company liked his work, *and* it gave him a better job.

It should be stressed that none of these types has any more inherent worth or intrinsic value than the other; each has a place in every paper. What the writer needs to understand is that each can be used to break a series of monotonous sentences; that each should correspond to the complexity of the concept—the more complex the concept, the more complex the form; that sentence form is a writer's—and the reader's— way of seeing the relative value of ideas; and therefore that the writer is responsible for seeing that the *least* complex form possible is used to present his ideas. This means that the responsibility for stressing some sentence elements and subordinating others is the writer's, not the reader's.

WRITING EFFECTIVE SENTENCES

Effective sentences result when the writer carefully examines the grammatical relationship of his concepts, when he attempts to transmit his ideas in the best possible manner for reader comprehension, and when he becomes conscious that the form of the idea is an aspect of its meaning. All of us have heard the expression "The tyranny of words," by which is presumably meant both the flexibility and the inflexibility of words, that they are always approaching but never reaching our meaning. But there is also a tyranny of form, for the relationship of concepts—aspects of sentence structure and the sentences themselves—indicates degrees of value and meaning to the reader. How often have we had to reply to someone, "But that's not exactly what I meant. What I meant was. . . ." "But," replies the questioner, "that is what you say here." Surely no one wants to defend the meaning of his ideas continually, yet how infrequent it is that writers, particularly those working under pressures that cause the inaccurate expression of concepts, ever attempt to consider the formal structure of their ideas. The following paragraphs suggest means for using form as an aspect of meaning.

Logical and grammatical importance

A clear and concise style results mainly from placing the *logical* subject and predicate of a concept in the *grammatical* subject and predicate positions. The proper relationship between important concepts and grammatical position is one of the most important guides to conceptual clarity. The reader needs to know which ideas are important, and structure performs this task. The most important grammatical positions in a sentence are the subject and the predicate of the main clause. We must ask ourselves, "What am I taking about?" or "What am I really trying to say?" or, in rhetorical terms, "What is the logical subject of my concept?" If, on examining the sentence, we find that the logical subject does not occupy the grammatical subject position in the main clause, and if what we predicate about that logical subject is not in the grammatical predicate position, then the sentence should be examined to determine if the disruption of normal order is useful or necessary. Here is a sentence in which lack of attention to form has obscured meaning:

> Movement is resisted by air in proportion to the surface of the moving object.

The grammatical subject is *movement,* but the logical subject is *resistance.* Restructuring the sentence with *resistance* as subject subordinates *movement* to its proper relationship as a modifier of the subject:

> Air resistance to movement is proportional to the surface of the moving object.

Here is another sentence in which the same problem occurs:

> The distance through which gases can diffuse is very small.

Here again the logical and grammatical subjects do not coincide. If we ask, "What are we really talking about?" we find we are really discussing the diffusion of gases, but the term *gas* is the subject of an adjective clause modifying *distance:*

> The distance *through which gases can diffuse. . . .*

The sentence gains clarity and emphasis if the main clause is rearranged to make the grammatical and logical subjects coincide:

> Gases can diffuse through small distances.

The following pairs of sentences represent changes which make the grammatical and logical subjects coincide. Examine them and be certain that you understand the necessity for the change and the clarity achieved in the second one of each group.

> This poetic drama was conceived in Byron's mind after he had been introduced to Goethe's *Faust*.
> Byron conceived this poetic drama after reading Goethe's *Faust*.

> The pressure we have observed in these experiments with air pressure is caused by trillions and trillions of molecules of air bouncing against the walls of a container.
> The air pressure observed in these experiments is caused by trillions and trillions of air molecules bouncing against the container walls. *Or:* Trillions and trillions of air molecules bouncing against the container walls caused the pressure observed in these experiments. (This sentence reorganization changes the passive *is caused* to the active *caused.*)

> The result of the experiment was a mixture of several brilliant colors: red, green, and blue.
> A mixture of three brilliant colors—red, green, and blue—resulted from the experiment. *Or:* The experiment resulted in a mixture of three brilliant colors, red, green, and blue.

In the majority of the preceding sentences, the grammatical subject has been changed to conform to the logical subject. However, the choice of a logical verb to describe the action and the positioning of that verb as the verb of the main clause are at least as important if not more so for conceptual clarity than is the choice of the logical subject. There are two primary causes for poor verbal sense: the logical verb often occurs as an adjective after the verb "to be," and the writer frequently chooses the most general verb from the group of verbs which might describe the action.

Improper Use of the Verb "To Be" and Other Linking Verbs
—(For a discussion of linking verbs, see the Glossary.) Someone who examined Shakespeare's plays to determine what con-
stituted their greatness once remarked on the almost total absence in his work of any form of the verb "to be." None of
us can hope to be as fine a writer as Shakespeare, but it is
possible to emulate this admirable trait to some degree. All
of us have heard some teen-ager exclaim something like, "He
was terribly handsome," and another reply, "He looked like
a god." Surely there is nothing wrong with this in conversation
(except possibly the exuberance), for if the first girl had said,
"This terribly handsome man looked like a god," the other
would have had nothing to add to the conversation. But writing is not conversation, and the writer should want to complete
his own thoughts as accurately as possible. Here are a few
sentences—each followed by a suggested change—in which
the verb "to be" or another linking verb is used inaccurately.

The amount of change *is* widely different in different
groups.
The amount of change *differs* widely in different groups.

Inherent variations *are* an occurrence which *is* quite
independent of any purpose on the part of the organism.
Inheritable variations *occur* quite independently of any
purpose on the organism's part.

He *felt* the need for a change.
He *needed* a change.

Writers should examine carefully all sentences that contain
some form of the verb "to be." Often the logical verb lurks
close by.

Abstract Rather than Concrete Verbs—Failure to choose
good verbs frequently results when the writer chooses the most
general verb describing an action, and then proceeds to add
modifiers to specify how the action was performed. But English is rich in verbs that describe an action specifically, and
there is no excuse for writing lazy and unemphatic sentences.
As an example, take the sentence, "The man hit the ball." The
verb *hit* describes only the general action; the manner of hitting is not described. A verb might have been chosen that
would have described the *manner* of hitting precisely: He
blooped, socked, dribbled, lined, lofted the ball.

The writer should be careful to choose verbs as specific as possible for the action. Consider the following sentences:

> The boy *walked* down the street with a limp.
> The boy *limped* down the street.
>
> The orator *talked* to the crowd.
> The orator *aroused* the crowd.
>
> The orator *excited* the crowd.
> The orator *electrified* the crowd.

Parallelism

In addition to making the logical and grammatical subjects and predicates coincide, writers should be certain that concepts that are similar in meaning should have similar gramatical form. If a statement begins with an infinitive, then other statements dealing with the same subject in a similar manner should begin with an infinitive, and so on:

> *To organize* the company's resources more fully and *to structure* the personnel more carefully, he suggested the company hire a consultant.

Parallelism is an effective device for utilizing grammatical form as an aspect of meaning, for readers encountering such "balanced" structures have a tendency to fasten on these elements and to grasp the meaning more easily than they would if the elements had been written separately. Using parallelism is habitual with good writers, but it is an acquired habit, and if you do not have the habit, it will be necessary to acquire it. The place to begin is in the paper on which you are currently working, and the time is now. Examine whatever you have written. If any sentence elements conform to the examples of faulty or nonexistent parallelism below, tne structures should be made parallel in the manner indicated.

> Faulty: Since *he was forced* to run the plant himself because of the lack of trained personnel, and *being* by nature a nervous man, the owner developed an ulcer.
>
> Correct: *Since he was* forced to run the plant because it lacked trained personnel and *since he was* a nervous man, the owner developed an ulcer.
>
> Faulty: He declined the offer because he knew *that* the plant lacked the proper equipment, and *he was*

sure that the company didn't have the proper personnel.

Correct: He declined the offer because he knew *that* the plant lacked the proper equipment and *that* the company didn't have the proper personnel.

Faulty: First he *wrote* a chapter of the paper. Then he *decided* to have lunch. Later he *edited* what he had written earlier.

Correct: First he *wrote* a chapter of the paper, then *decided* to have lunch, and later *edited* what he had written.

Faulty: *Hoping* to be hired and *to make* an impression, the engineer bought a new car.

Correct: *Hoping to be* hired and *wishing to make* an impression, the engineer bought a new car.

Subordination and Coordination

Both these terms have been mentioned earlier, but both will be treated here as means toward sentence style rather than as sentence elements. Coordination is the means by which independent units of meaning of similar importance are linked as equals; subordination is the means by which units of less importance are subordinated to more important ones. It is obvious that in such sentences as "He felt tired. He left the class," one of the sentences is less important than the other and that therefore the less important element is subordinated: "Since he felt tired, he left the class." "Feeling tired, he left the class." Not all sentences are as simple as these, and the question of which element to subordinate is one that the writer has to answer on the basis of reader interest, main thesis, and other such considerations. Consider the following:

Wöhler had shown that the synthesis of organic compounds can come by way of the living cell or from lifeless contents of the test tube.

As Wöhler had shown, the synthesis of organic compounds can come by way of the living cell or from the lifeless contents of the test tube.

In the first sentence the writer wanted to emphasize Wöhler, and so "synthesis" is in a dependent clause. In the second sentence the writer wanted to emphasize the contribution, so Wohler is in the dependent clause.

Proper subordination, therefore, results when an author has thought through his material. At times, however, material that should properly have been subordinated is left as coordinate because of stylistic considerations. In the following examples of proper subordination, any sentence might have been made coordinate or allowed to stand as two separate sentences. The last example illustrates coordination. The decision depends on the surrounding sentences and the tendency of the material.

The DNA enters the bacterium. It disrupts the cell's normal working.

When the DNA enters the bacterium, it disrupts the cell's normal working.

The motor idled beautifully. The car was a 1948 Cadillac.

The motor of the 1948 Cadillac idled beautifully.

Bacteria all look very much alike. They can be classified into a few basic types.

Since bacteria all look very much alike, they can be classified into a few basic types.

The motor worked well. The suspension system needed attention.

The motor worked well, but the suspension system needed attention.

SOME COMMON SENTENCE FAULTS

Misplaced modifiers

A misplaced modifier is a word, phrase, or clause not clearly related to some sentence element. Because English is a positional language, it is relatively simple to make mistakes in modification. However, if the writer adopts the habit of placing his modifiers as close as possible to the word they modify, he will run into this problem only infrequently. In the following sentence the unusual placement of "research-oriented" confuses the reader:

The company hired three physicists, research-oriented.

Who was research-oriented, the company or the physicists or both? The writer should let the reader in on the secret:

The company, research-oriented, hired three physicists.
The research-oriented company hired three physicists.
The company hired three research-oriented physicists.

The acid burned his finger which was odorless and colorless.
The odorless and colorless acid burned his finger.

Unnecessary shifts of voice or subject

Avoid writing sentences that shift from the active to the passive. Such shifts are awkward, and the use of the passive requires a shift in the subject:

Faulty: When *I found* the trouble in the motor, the *repair was begun.*

Correct: When *I found* the trouble in the motor, *I began* the repair.

Faulty: The *mold became* so expensive that *it could not be supported* by the company.

Correct: The *mold became* so expensive that the *company could not support* it.

Faulty: The blast furnace *could be heard* roaring as we approached.

Correct: *We could hear* the blast furnace roaring as we approached.

Unnecessary shifts of person or number

Faulty: When *you* have an excellent staff, *one* should feel fortunate.

Correct: When *you* have an excellent staff, *you* should feel fortunate.

Correct: When *one* has an excellent staff, *one* should feel fortunate.

Faulty: *We* could find only one fault with his equation: when *you* worked it out *you* always got a different answer.

Correct: *We* could find only one fault with his equation: when *we* worked it out *we* always got a different answer.

Sentence fragments

Every sentence should have a subject and predicate and should state a complete thought. A group of words written as a sentence that does not have these two elements is called a sentence fragment. Frequently sentence fragments result when the writer places his final punctuation too soon:

> I had expected the scientist to be old and shy, but actually he was young and outgoing. *With an intense inquiring mind and an interest in sports.*
>
> The entire laboratory was geared to the effects of cold on various metals. *With several experts on the staff.*
>
> He was one of the most active men I have ever met. *Racing from his house to the lab, and from the lab to the administration building.*

Sentence fragments are frequently used by writers to achieve emphasis. There is nothing wrong with using fragments intentionally. But the accidental use of sentence fragments is a sign of amateurish writing.

CHAPTER 10

THE RIGHT WORD

DICTION

Correct diction is the basic element in all writing. Words have to be well chosen, for precision increases clarity and interest. Good diction means the absence of ambiguity, obscurity, and misunderstanding.

General words, unlike some scientific ones, have more than one meaning and more than one quality. Most words do not simply *denote* (the meaning found in the dictionary) something, as the symbol H_2O denotes water. Words also *connote*—they *imply* meanings in addition to the denoted meaning. Many words have similar denotations, but different connotations; for example, we have many words meaning dog. Consider these: canine, cur, mutt, mongrel. It is quite obvious that "mutt," although it means a kind of dog, connotes much more to the reader than simply a dog of undetermined lineage. "Canine" is much more formal and also much less visual and concrete than "mutt"—it is less connotative. The situation in which we would choose one of these words for "dog" would be determined by the degree of formality in a paper. Words have different degrees of appropriateness to different writing situations. Writers who wish to use words precisely have the responsibility of considering all aspects of a word.

Use of the specific rather than the general word

The admonition to all good writers is: always prefer the concrete to the abstract, the specific to the general. The general word refers to the class or group, the specific to a member of the group. "House" is a general word that refers to the group; "cabin" is specific. The specific word "cabin" permits the reader to visualize the object, as "house" does not. The specific enables the reader to comprehend more exactly and more readily.

Use of the concrete rather than the abstract word

Abstract words refer to ideas and qualities. Concrete words refer to particular objects and particular actions. "Automobile," "buggy," and "chariot" are concrete; "mobility," "speed," and "stateliness"—qualities that might be associated with these concrete words—are abstract. Concrete words represent our concepts directly and specifically. Technical writers should use the concrete and specific word. Consider the following example:

> As cells grow, troubles arise. The rate at which the sustenance of life enters this elemental unit and the useless by-product of that energy-producing material departs depends upon the total exterior surface of the unit under discussion.

Perhaps no one actually writes *exactly* like the author of these sentences, but some authors come very close. Examine your own writing and the example above. If your writing is full of abstract terms, make them concrete. These sentences should have been written:

> As cells grow, troubles arise. The rate at which *food* enters the *cell* and *waste* departs depends upon the *cell's surface area.*

The value of any piece of writing depends ultimately upon the writer's word sense; every writer should cultivate the habit of thinking about words. Here are some sentences in which abstract and general words have yielded flat, stereotyped writing. Each sentence has been rewritten using specific words.

> The architect received a fee for aiding in the building's design.
> The architect received a fee for aiding in the *library's,* the *temple's,* or the *department store's* design.

> A long-sustained production increase at the present rate will accomplish wonders in Latin America.
> A long-sustained production increase of 5% will *industrialize* Latin America in *ten years.*

Great changes *came over* all the sciences of living in the years between 1830 and 1880.
Great changes *developed* in all the sciences of living things between 1830 and 1880. (Note the omission of the non-functional "in the years.")

In this way two early scientists *made the discovery* that all materials had electric properties under the right conditions.
In this way two early scientists *discovered* that all materials had electric properties under the right conditions.

Like others, he *got* some electric apparatus for amusement.
Like others, he *obtained* some electric apparatus for amusement. (Even in this revision, the manner of obtaining is not specified; a better verb, such as *borrowed, bought,* or *constructed,* indicating the manner of obtaining, would be preferred.)

Seeing this, he *made* the invention of the "lightning rod."
Seeing this, he *invented* the "lightning rod."

The use of metaphor

The metaphoric use of language provides the writer with a technique that allows him to employ words because they have some analogy to the concepts he is discussing. Metaphors are used when the writer recognizes a similarity (physical resemblance, similarity of function, or the like) between two essentially dissimilar ideas. He *flew* down the street; He *shot* into the laboratory.

We write metaphorically whenever words are so used that a quality or function belonging to one thing is transferred to another. Language is constantly enriched by the general acceptance of metaphoric usage as a normal means of dealing with the concept: the *bed* of a stream, a railroad *bed*. Good writers constantly enliven their work with new, vibrant metaphors. For example, the title of a popular biological work is partly metaphoric: *The Wellsprings of Life.* This title is much more vivid than *The Source of Life.*

There is no reason why technical writers should avoid metaphoric language, but its use should be careful and considered. Be certain that the metaphor is new and alive; worn-out metaphors—ones that have been in the language for a long time—

are no longer metaphors. In many cases they are simply dead terms.

Do not use metaphoric language too frequently in a paper. The reader's attention will be directed to the language rather than to the ideas. Use metaphors when you think the writing is flat and uninteresting; use them as stylistic devices. Get the habit of spicing your writing with a dash of metaphors. Here are some possible metaphors for the sentence "The motor worked well":

> The motor *purred*.
> The motor *sang*.
> The motor *hummed*.

Here are some for the opposite:

> The motor *balked*.
> The motor *coughed*.
> The motor *jerked* the car along.

WORD LEVEL

Words can be colloquial, informal, and formal. Words, like ideas, have contexts that are appropiate for their use. The most general language is colloquial, used for everyday speech. Informal language, used for personal letters, office memoranda, or informal essays and speeches, is a refinement of colloquial. Formal language is the most restricted language level. Words that are acceptable in colloquial or informal language are not acceptable in formal writing. The writer should be certain that the word level is appropriate to the general tone of the paper. "Nag" might be all right in colloquial or informal writing, but "horse" would have to be used in formal writing. Writers should develop a sensitivity to words. Both the dictionary and the thesaurus will help.

The use of the dictionary

The dictionary is the primary tool for word accuracy. It should be consulted whenever any uncertainty arises over the use, meaning, or spelling of a term. When consulting a dictionary, be certain to examine all the meanings listed for an entry, not simply the first one that strikes the eye. Examine the derivation, which is sometimes helpful in fixing precise meaning and in understanding how the word has extended its meaning. Pay attention to the suggested synonyms. Often the examination of the synonyms will lead the writer to a word of more precise meaning than the word he is looking up. The dictionary is the writer's basic tool—it should be used intelligently. Here is a typical dictionary entry:

> **obe·di·ent** \-ənt\ *adj* [ME, fr. OF, fr. L *oboedient-, oboediens,* fr. prp. of *oboedire* to obey — more at OBEY] **:** submissive to the restraint or command of authority — **obe·di·ent·ly** *adv*
> **syn** DOCILE, TRACTABLE, AMENABLE: OBEDIENT implies compliance with the demands or requests of one in authority; DOCILE implies a predisposition to submit to control or guidance; TRACTABLE suggests having a character that permits easy handling or managing; AMENABLE suggests a willingness to yield to demands, advice, or contrary suggestion

> By permission. From Webster's Seventh New Collegiate Dictionary, copyright, 1963 by G. & C. Merriam Co., publishers of the Merriam-Webster Dictionaries.

Notice the information we are given. The word is an adjective (*adj.*). The hyphen before -*ent* refers us to the entry above for pronunciation; we are told how to break into syllables. The etymology is traced: the word entered English from Old French (OF), which obtained the word from Latin. Then we are given the meaning and the adverbial form. After that, several synonyms are suggested, and distinctions in meaning for each of the synonyms are discussed.

Surely any source which tells so much about a word should be consulted frequently!

Use of the thesaurus

All good writers are familiar with Roget's Thesaurus, a book that suggests areas of meaning in which words normally function. This thesaurus, and other similar works, such as *Soule's Dictionary of English Synonyms,* should be consulted when the writer wants to find a more precise term for a word. Works of this sort are also extremely helpful in determining subtle differences in words of similar meaning. Here is an example from Soule:

> **Obedient,** *a.* Compliant, submissive, deferential, dutiful, duteous, respectful, yielding compliance, observant, regardful, subservient.

Webster's Dictionary of Synonyms provides brief definitions to help the writer make his choice:

> **obedient.**   Obedient, docile, tractable, **amenable,**
> **biddable** are synonyms carrying the common meaning of
> submissive to the will, guidance, or control of another.
> Though applied chiefly to persons, they are by extension
> applicable also to things. **Obedient** implies due compli-
> ance with the commands or requests of a person or power
> whose authority one recognizes or accepts; as, *obedient*
> to the law; children trained to be *obedient* to their par-
> ents. "Exhort servants to be *obedient* unto their own
> masters" (*Titus* ii. 9). When applied to things, it implies
> compulsion by a superior force, movement in accordance
> with natural law, or the like; as, tides *obedient* to the
> moon. "And floating straight, *obedient* to the stream,
> Was carried towards Corinth" (*Shak.*). **Docile** implies a
> responsiveness to teaching, but it stresses either a . . .

It should be obvious from the two selections above that these two works are invaluable for a writer seeking either to extend his vocabulary or to understand the precise meaning of a term. Both works should be on the desk of every writer. They should be consulted frequently.

CHAPTER 11

VERBS

Verbs express action, condition, or being.

> The man *hit* the ball.
> The ball *was* round.
> The man *is* a doctor.

PRINCIPAL PARTS

There are three principal parts (or forms) of a verb: the simple form (the first form in the dictionary), the past, and the past participle. If we know these forms, we can formulate all the tenses of a verb.

Verbs are both regular—those that form their past and past participle by adding *-ed, -d,* or *-t* to the base form—and irregular. Irregular verbs form the past and past participle by a vowel change in the simple form.

Stem	*Past*	*Past Participle*
	REGULAR	
wish	wished	wished
look	looked	looked
rake	raked	raked
	IRREGULAR	
run	ran	run
go	went	gone
sing	sang	sung

TENSE

Tense is the term used for the different forms verbs assume when making time distinctions, and in general these terms indicate the time at which the action occurred. The principal tenses in English are: present, present perfect (sometimes sim-

119

ply called the perfect), past, past perfect, future, and future perfect.

> PAST PERFECT—action prior to another past action.
> He *had* already *left* the laboratory when I arrived.
> PAST—action begun and completed in the past.
> He *worked* in the laboratory for ten years. (He no longer works in the laboratory.)
> PRESENT PERFECT—action beginning in the past and continuing to the present.
> He *has worked* in the laboratory for ten years. (He is still working in the laboratory.)
> PRESENT—present and habitual actions, simple future, conditions true for all time.
> I *walk* ten miles every day.
> I *leave* tomorrow.
> Water *freezes*.
> He *runs* well.
> FUTURE—action begun in the present and continued in the future.
> He *will leave* the laboratory when I finish the project.
> FUTURE PERFECT—action which will occur prior to some future action.
> He *will have left* the laboratory by the time the project is completed.

Progressive tenses

All the tenses in English have a progressive form that intensifies the verbal action. The progressive forms are being increasingly used, and writers should remember that they exist:

> Present: I *am walking,* he *is walking.*
> Present perfect: He *has been working* in the laboratory for ten years.
> Past: He *was working* when I arrived.
> Past perfect: He *had* already *been working* for the company ten years when I arrived.
> Future: He *will be leaving* for Europe at the end of summer.
> Future perfect: He *will have been working* for the company for ten years by the time the project is completed.

Sequence of tenses

In general, writers have the greatest difficulty with tenses in structuring the temporal relationship of actions that occur at different times in the past. In many instances in nonformal writing, the past perfect is no longer used to distinguish past temporal relationships, and both the past action and the action that occurred prior to the past action are described with the simple past. However, before a writer violates or decides to forego a temporal distinction, he should be aware that he is doing so. Nothing is more indicative of poor writing than the *accidental* violation of normal grammatical relationships. Moreover, in many cases the violation of these relationships results in ambiguity and confusion. Consider:

> The policeman *arrested* the criminal who *escaped.*

Did the criminal escape after the policeman arrested him? The reader is confused because he has not been informed of the temporal relationship of the actions. Such confusion can be overcome by the use of the proper sequence of tenses:

> The policeman arrested the criminal who *had escaped.*

Notice the clarity achieved by paying attention to the correct temporal relationships in the following sentences:

> The next great step was to prove that voltaic electricity was, as Fabroni *surmised,* itself the product of chemical action.
> The next great step was to prove that voltaic electricity was, as Fabroni *had surmised,* itself the product of chemical action.

> The trap that *caught* the bear resembled the jaw of a boa constrictor.
> The trap that *had caught* the bear resembled the jaw of a boa constrictor.

> When the abstract reached his office, the supervisor *left.*
> When the abstract reached his office, the supervisor *had left.*

ACTIVE AND PASSIVE VOICE

Verbs in the active voice transmit action from the subject to the object:

The man *hit* the ball.

Verbs in the passive voice do not transmit action. The subject is not the actor, but is *acted upon:*

The ball *was hit* by the man.

Whenever possible, verbs should be active, not only because the subject performs the action, but also because passive verbs obscure the verbal action and the logical subject.

Reports suffer more from the passive complex than from any other grammatical problem. Give a poor writer an opportunity for an emphatic statement and he will avoid it every time—frequently by using a passive rather than an active verb. The passive should not be used in formal writing if the active can be used instead. The conscious use of the passive, however, can sometimes be very effective as a stylistic device. What most editors and grammarians decry is the unconscious use of the passive. Let us see what happens grammatically when we transpose a sentence from the active to the passive:

Passive: The criminal *was caught* by a policeman.
Active: The policeman *caught* the criminal.

The passive voice requires more words to state a concept than the active, the logical subject becomes the object of a preposition, and the normal object becomes the subject.

It should be evident, since the normal object becomes the subject of a passive verb, that one of the principal uses of the passive is the emphasis of the normal object of an active sentence.

MOOD

Based on patterns prevalent in many European languages, verbs in English were divided into three moods: indicative, imperative, and subjunctive. The indicative is used for statements of objective fact. The imperative mood—usually associated with commands—and the subjunctive—with wishing, desire, and conditional statements contrary to fact—were utilized to explain verbal constructions that differed from the normal. However, the continued use of such terms as "imperative" and "subjunctive mood" is confusing rather than clarifying.

Terminology should be functional. To say that "Stop, thief!" is in the imperative mood is simply to recognize that with commands and requests the simple form of the verb is used.

Similarly with the subjunctive, now confined almost exclusively to questions of style in formal writing, it is possible to say that, since there are so few forms of the subjunctive in current use, their use is idiomatic.

The principal uses of the subjunctive are two. One is with conditions contrary to fact:

> If I *were* a scientist, I would spend some time explaining the advantages of a scientific career to students.

The other occurs in "that" clauses following resolutions, demands, recommendations, and suggestions:

> We ask that the prisoner *be* hanged.
> He recommended that I *be* informed.
> I suggested that he *be* informed of the outcome of the experiment.

In most cases, however, standard or general English would avoid the subjunctive in two of the three sentences above:

> He recommended *that they inform me*.
> I suggested *that they inform him* of the outcome of the experiment.

SEQUENCE OF TENSES IN CONDITIONAL SENTENCES

Conditional sentences are composed of two elements—a conditional clause (usually introduced by "if") and a main clause, whose validity rests upon the condition stated in the "if" clause. There are two forms of conditional sentences, those in which the condition is possible and those in which it is not. In both forms there is a sequential relationship between the verbs in the conditional clause and in the main statement.

POSSIBLE CONDITIONS

present tense in "if" clause	*present or future in main clause*
If you *go,*	he *goes* too.
If you *go,*	he *will go* too.

past tense in "if" clause	*"would" and simple form of verb in main clause*
If he *showed* his cards,	I *would show* mine.
If he *raised* his offer,	I *would raise* mine.

NEGATIVE CONDITIONS

past tense in "if" clause	*"would" and simple form of verb in main clause*
If you *were* king,	I *would be* prime minister.
past perfect in "if" clause	*present perfect and "would" in main clause*
If he *had raised* his offer,	I *would have raised* mine

VERBALS

Verb forms that function as modifiers or nouns are called *verbals*. They are the infinitive, the present and past participle, and the gerund. Verbals have in common the fact that they are modified adverbially (the gerund, being a noun, may also be modified adjectivally) and that they can take objects. The gerund and the present participle are similar in form, both ending in "-ing," but the gerund and the participle are distinguished by their function—the gerund is a noun and the participle an adjective.

participle: *Walking* through the room, the women talked of art and politics. (The phrase *Walking . . .* modifies *women*.)

gerund: *Swimming* is fun. (*Swimming* is the subject.)

¶ *Gerunds* can be used effectively to achieve economy and emphasis in certain frequently occurring phrases, such as "in the . . . of":

In the revision of a report, style should be considered.
When *revising* a report, consider style.

His chief recreation is *the tying of* flies.
His chief recreation is *tying* flies.

The telling of stories is a favorite recreation.
Telling stories is a favorite recreation.

Remember that gerunds are nouns, and are frequently modified by possessive adjectives:

Faulty: I would not consider *him* leaving. (*Him* is not the object of consider, but modifies *leaving*.)
Correct: I would not consider *his* leaving.

¶ *Participial phrases* are used to modify nouns. They function exactly like adjectives and they occasionally dangle:

Running through the room, the door slammed in my face.

Since introductory participial phrases modify the subject, the *door* is doing the running. While there is no ambiguity and perhaps some amusement in a sentence of this sort, such misplaced and dangling modifiers offend and often distress readers.

Faulty: Being concerned with the effect of temperature on the structure of the lead, several crystals interested me.

Correct: Because I was concerned with the effect of temperature on the structure of the lead, several crystals interested me.

Correct: Being concerned with the effect of temperature on the structure of the lead, I was interested in several crystals.

¶ The *infinitive* is the basic form of the verb: *to swim, to run.* *To* is often omitted, especially after *shall, will, may, must* and other auxiliaries: He *may* run.

CHAPTER 12

PUNCTUATION

THE COMMA

Restrictive and nonrestrictive

Modifiers—Much of the difficulty of punctuating an English sentence can be overcome if the distinction between restrictive and nonrestrictive modifiers is understood. A restrictive modifier is one that *cannot* be omitted from the sentence without altering the meaning; it is necessary for comprehension. A nonrestrictive modifier is one that *can* be omitted without altering the meaning. Nonrestrictive modifiers are cut off by commas; restrictive modifiers require no punctuation:

Men seldom make passes at girls *who wear glasses.*

Since it is quite obvious that the term *girls* requires some sort of modification to identify which particular group of girls men don't make passes at, the modifying clause *who wear glasses* is restrictive. The sentence makes no sense without it. Therefore the sentence requires no punctuation.

Men always make passes at girls *no matter what they look like.*

All girls are meant here; therefore, the modifier does not restrict or qualify girls in any way. The sentence should have a comma added before *no.*

My uncle *who lives in Hoboken* is an engineer.

Before a sentence such as this one can be punctuated, the writer must determine whether there is more than one uncle. If there is, then the clause is restrictive, because it identifies which uncle is the engineer; if, on the other hand, there is only one uncle, then the term *uncle* needs no qualification—there can be no mistake about which person is intended. In this case the qualifying statement is simply additional, non-restrictive information, and needs commas before *who* and after *Hoboken.*

Following are several sets of sentences; for the first few sentences the reasons for the punctuation are given, for the others the reader should apply the principles of restrictive and nonrestrictive to be certain he understands them.

> My father, who is a management consultant, is away on a trip.
> The man who is a management consultant has a fine future.

In the first sentence the term *father* identifies the person adequately, and the modifying clause, while informative, is really unnecessary. In the second, the clause is necessary to identify which man is intended. The clause is necessary and restrictive.

> Albert Einstein, who was one of the greatest physicists and mathematicians, lived for a time in Princeton.

Since the name *Albert Einstein* completely identifies the subject of the sentence, the material between the commas is added information and is nonrestrictive. The sentence would make sense without it.

> Several physicists who have made worthwhile contributions to scientific thought were honored at the dinner.

Since the clause *who have made worthwhile contributions to scientific thought* identifies which physicists were honored, the clause is restrictive.

Here are a few sentences on which you can test your comprehension of the principles involved in restrictive and nonrestrictive. Identify the modifying element and determine whether commas are necessary:

1. Londonderry House which is one of Mayfair's last remaining stately homes was sold at auction today.
2. Galvani who was an Italian expert on anatomy then made public his working out of this chance discovery.
3. Each stratum had its own characteristic fossils which did not appear in other strata.
4. Shallow rock pools which are present in many areas of the northeast coast are the nursery of tiny shore crabs.

5. Pools formed in shallow rock are the nursery of tiny shore crabs.*

Appositive—An appositive is a noun, or words used as a noun, that renames the noun immediately before it. Appositives may be either restrictive or nonrestrictive, according to whether they are necessary or unnecessary to identify the word they restate. Most appositives are nonrestrictive.

Restrictive:	William of Orange (there are many Williams)
	Richard the Lion Hearted (many Richards)
	John the butcher
	My uncle John (more than one uncle)
Nonrestrictive:	Richard the Lion Hearted, King of England
	John, *our* butcher
	My uncle, John (only one uncle)

Parenthetic Expression—An expression inserted in a sentence that interrupts the thought of the sentence is a parenthetic expression. All are nonrestrictive.

Genes, *for example,* cannot always be clearly differentiated as dominant or recessive.
He was, by the way, one of my best friends.
In Britain, for instance, the countryman has two chances of seeing the effect of animal light.

Coordinate clauses

Probably nothing so simple causes so much confusion as the punctuation of coordinate clauses. A coordinate sentence consists of two or more independent sentences joined either by a comma and a coordinating conjunction, or by a semi-colon (see below, p. 131).

There are seven common coordinating conjunctions: *and, so, or, for, nor, yet,* and *but.* They should be memorized. To join two independent clauses, place a comma where the period would normally occur and add one of the coordinating conjunctions:

---

*   1. N,   2. N,   3. R,   4. N,   5. R

The man hit the ball. He ran to first base.
The man hit the ball, *and* he ran to first base.
The man hit the ball, *so* he ran to first base.
The man missed the ball, *but* he ran to first base.
The man missed the ball, *yet* he ran to first base.
The man missed the ball, *for* he had swung too high.

While is sometimes used as a coordinating conjunction:

I singled twice, *while* he had only one hit in four times at bat.

"And" used as a compounding conjunction

Everyone is familiar with the use of *and* in such phrases as "bacon *and* eggs" or "Jack *and* Jill," and no one would be inclined to punctuate phrases such as these, but when similar grammatical forms are more complex, problems do occur. These principal uses of *and* as a compounding conjunction do not take a comma before *and;* following:

Compound Verbs—No punctuation is needed between verbs describing similar actions when they have a common subject:

He *poured* the milk and *mixed* the batter.
He *entrusted* the formula to a friend and *persuaded* him not to reveal the contents.

Compound Subjects and Objects—No punctuation is needed between subjects or objects when they relate to the same verb:

Trimming the hedges and *cutting* the lawn were his week-end activities.
They toured the *lakes* in a car and the *mountain* in a bus.
He said *that he would tour Mexico* and *that he would buy me a piece of Mexican silver.*

When *but* and *for* are used as prepositions, no punctuation is necessary unless the phrase is nonrestrictive.

Introductory phrases and clauses

Although the rule for separating introductory phrases is rather vague (a comma is used if the phrase is long), the writer will never be wrong if he uses a comma after all introductory phrases of more than three words:

no comma   In the end he returned.
Because of this he determined to go.

comma   From a distance of two or three miles, the
               men could barely contact the ship.
               Realizing the seriousness of the situation, the
               president summoned all unit heads to a
               conference.

All verbal phrases are separated from the sentence by a comma:

> Striding from the room, he turned left and fled.
> Startled by the sounds, the deer bolted.
> Startled, the deer bolted.

Coordinate adjectives

Adjectives are coordinate when they modify a noun independently—when the relationship of each adjective to the noun is similar. Whenever this occurs, the comma is used.

> The *cool, clear, sparkling* water spilled from the *old oaken* bucket.

Cool, clear, and *sparkling* all modify *water* in a similar manner. Each is an independent modifier of the noun—the cool water, the clear water, the sparkling water. However, *old* not only modifies bucket, but also the material of which it is composed. *Old* and *oaken,* therefore, are not separated by commas. A good test to determine if the adjectives should or should not be punctuated is to put *and* between pairs of them. If the adjectives on both sides of *and* are similar, then a comma is used between them:

> bright and sparkling, cool and clear

but not

> old and oaken

Another good test is to attempt to change the order of the adjectives. If the order cannot be changed without changing the meaning, the adjectives are not coordinate.

Subordinate phrases and clauses

Use a comma to punctuate subordinate phrases and clauses that explain, amplify, or contrast with the main clause if they are not closely related to the main clause:

He was in favor of expanding the program, although he objected to the use of capital funds to finance it.

The spring-loaded top pivot assures constant, taut door-track engagement, which is one of the features of the product.

Other uses of the comma

The comma also separates the day of the month from the year: May 13, 1921. It is often not used to separate the month from the year: September 1921. It also separates cities and villages from states: Greenwich, Connecticut; Ramapo, New York.

The comma follows the salutation in informal letters: Dear John, . . .

The comma separates degrees and titles from names: Charles Wallace, A.B.

THE SEMI-COLON

Use the semi-colon 1) to punctuate coordinate sentences, and 2) to separate elements in a list or series when any of the elements contains a comma.

¶ The semi-colon is most frequently used to replace the comma and coordinating conjunction between *coordinate sentences:*

He was one of the most competent men on the staff; I was very sorry to lose him.

She preferred to be busy around the house; I preferred to rest and read.

¶ The semi-colon is used to separate the elements of an internally punctuated list or series:

She bought a yellow, brown, and green hat; a wide-brimmed hat; and, of course, a pair of leather gloves.

He bought Joyce's *Ulysses* (Paris, 1926); O'Casey's *Red Roses for Me* (London, 1960); and Synge's *Playboy* (Dublin, 1909).

THE COLON

Use the colon to indicate that a series, an explanation, or a restatement follows. It sometimes takes the place of the awkward phrase *such as:*

He had many qualities in his favor: integrity, vitality, and knowledge.

The most curious fact about the transaction was this: he was completely unaware of the object's value.

The colon is also used:

- After the salutation in formal letters: Dear Sir: . . .
- To separate numbers in equations and proportions: The proportions are 5:10:5; to separate hours and minutes: 12:00; 2:45.
- To introduce a long quotation which will be separated from the text by indenting and by using single space: The full text of his remarks follows: . . .

THE DASH

The dash has two principal uses. It appears in place of the comma when the nonrestrictive material is far removed from the principal meaning of the sentence:

Thus, the report by the Harvard team seems also to add new weight to the hope shared by researchers all over the world: that a substance—like the one Dr. Fogh and Dr. Allen believe they are working with—will be found that can be used specifically against cancer.

The dash also separates a series or list from the rest of the sentence when the series will be wholly contained within the sentence. (A colon is used if the series terminates the sentence. See the example above.)

He was so affected by the environment—social, political, and economic—that he left the country.

He was affected by the entire environment: social, political, and economic.

Note that the dash separates material from the sentence more sharply than the comma, but not as sharply as parentheses.

PARENTHESES

Parentheses have two uses. They enclose material interpolated into a sentence that is far removed from the meaning of the sentence; that supplies information the reader probably knows but might not recall:

Researchers reported (Chicago, 1962) that there is almost general agreement that the genetic code of life is universal.

Parentheses also enclose the place and date of publication both in the text and in footnotes. (See example under "The semi-colon," p. 131.)

QUOTATION MARKS

¶ Quotation marks indicate material taken from another writer. If the quotation is less than five lines, it is usually contained in the paragraph and enclosed in quotes. If it is more than five lines, no quotation marks are used—it is separated from the paragraph by indenting and changing to single space, or by changing the type face. When the quotation is separated from the text in this manner, the last word in the paragraph is followed by a colon. End punctuation (period, question mark, and exclamation mark) comes within the quotation. The semi-colon and colon are placed outside the quotation marks.

¶ Quotation marks indicate dialogue:

> He said, "I will be along in a moment."
> "I will be along in a moment," he said.

When the speaker's name interrupts the quotation, no capital is used when it is resumed:

> "I shall be along," he said, "in a moment."

¶ Quotation marks are used in bibliographic references to separate titles of articles, poems, and short stories from the different titles of larger works in which they appear.

HYPHEN

The following rules for hyphenation are based on those in *Webster's Seventh New Collegiate Dictionary** and the United States Government Printing Office *Style Manual*.†

A hyphen is used:

¶ To *avoid doubling* a vowel or tripling a consonant (except after *co, de, pre, pro, re*).

thimble-eye   shell-like   micro-organism

---

* G. & C. Merriam Co. (Springfield, Mass.), 1963.
† Superintendent of Documents, Washington 25, D.C. Consult this manual for an exhaustive list of hyphenated words.

But *re* and *pre* are hyphenated if the following root word begins with *the same* vowel:

> re-elect   pre-eminent   re-enter

¶ To *prevent mispronunciation* or to insure a definite accent on both compound elements:

> dynamo-electric   contra-indicated   re-treat (treat again)   re-cover (cover again)

¶ To *join a letter or a numeral to a noun or participle,* when this compound is used as an adjective:

> U-boat   T-shirt   50-watt   330-horsepower
> X-ray   T-shaped

but:

> T square   L head

¶ To *join the elements* of an improvised compound:

> make-believe (*n*)   blue-pencil (*v*)

¶ In *compounds naming the same person or thing under a double title* or two aspects:

> secretary-treasurer   treasurer-manager

¶ In all *compound numerals* between twenty-one and ninety-nine, in technical *compound units of measure,* and in *complex compass terms:*

> twenty-eight   light-year   horsepower-hour
> north-northeast   6-footer   one hundred and twenty-eight

¶ In a *compound containing an apostrophe* in one of of its elements:

> bull's-eye   mare's-nest

¶ In a *compound predicate adjective,* the second element of which is a past participle:

> bull-necked   pig-headed   bell-shaped
> air-minded   wind-blown   full-bodied

¶ In *compound numerals and fractions* used as adjectives:

two-thirds majority   two-thousandths tolerance

but:

a majority of two thirds   a tolerance of two thousandths

¶ In a *series of compound modifiers* which follow a noun:

vases: 1-inch, 6-inch, 4-inch
alternating current motors: ½-horse, ¼-horse, 2-horse

¶ In *compounds with a common basic element*:

a six- or eight-cylinder motor
8-, 12-, and 16-foot boards

¶ In *fractions,* unless the hyphen appears in either the numerator or the denominator:

two-thirds   forty-four fiftieths
thirty-three thirty-seconds

¶ To indicate a *word division* at the end of a line of type or print; however, most editors prefer as few divided words as possible. Divide words only when absolutely necessary. In general, words are divided between syllables (when in doubt about syllable division, consult a dictionary):

ad-min-is-ter   catch-ing   win-ner

Do not divide words of one syllable, such as *sad, lone.* Do not divide two-syllable words beginning or ending with single

letters: *e*-nough, man-*y, a*-mong.

Words already hyphenated (*foot-candle, volt-ampere*) should be divided at the hyphen.

APOSTROPHE

The apostrophe is used primarily to indicate the possessive case and to indicate the omission of a letter or letters in contracted words:

Possesive singular: boy's, girl's, horse's, motor's,
                     another's

Possesive plural: boys', girls', horses', motors'

Omissions: don't, can't, he'll

Its is the possessive form of *it* (remember *he* and *his*). *It's* is the contracted form of *it is*. There is no word *its'*.

CAPITALIZATION

The following rules for capitalization are intended to cover the principal occasions. They are not all-inclusive. For a complete guide, see the GPO *Style Manual*, mentioned above, p. 133, or consult a good recent dictionary. Capitalize:

¶ The *first word* of a sentence, the first word of a formal quote:

> He said, "Our work has just begun on this project."

¶ *Proper nouns* and derivatives of proper nouns, when used as proper nouns:

> Victoria, Rome, Spain, Morocco
> Victorian customs, Roman laws. *But:* morocco binding, brussels sprouts, plaster of paris

¶ *Common nouns and adjectives in proper names:*

> Massachusetts Avenue. *But:* the avenue
> Cape of Good Hope. *But:* the cape

If the *shortened form* of a proper noun is commonly understood to stand for the noun, it is capitalized:

> The District (District of Columbia)
> The Lakes (The Great Lakes)

¶ *Individual names,* official and government titles, names of corporations, organizations, societies, and companies:

> John Doe, Henry James, Linnaeus.
> President, Governor, Senator, the Constitution, Federal Reserve Bank, Federal Communications Commission, National Recovery Act.
> Masons, Socialists, Bakelam Company,
> Mechanical Engineering Society.

¶ *Geographical terms* when combined with other terms to indicate specific proper names:

> Grand Canyon, Ramapo River, Niagara Falls, Lake Erie, Japanese Current. *But:* Peruvian mountains, South Pacific islands

¶ *Months of the year, days of the week, holidays and holydays:*

January, Wednesday, Columbus Day, Hanukkah, Good Friday

¶ Names of *geological eras,* periods, epochs, strata:

Neolithic, Jurassic, Stone Age

¶ Names of *genera* but not of species; Latin names of classes, families, and other groups above species:

Tetrapoda, Reptilia, Utheria, Gastropoda, Homo sapiens

¶ Names of *countries,* peoples, regions, localities and geographic features:

Philippine Islands; Hindus; Long Island Sound; Schenectady, New York; the Far East; the East; the Orient; the West; the Eastern Shore (Chesapeake Bay)

¶ *Trade names:*

Ivory soap, Ovaltine. But: castile soap, german silver.

¶ First letters of words in titles of *books, periodicals, and poems,* except in articles, prepositions, and conjunctions:

Shakespeare's *Hamlet,* "DNA as a Factor in Heredity," *Journal of the American Medical Association*

¶ Titles of *reports, technical articles,* and *proposals.* Capitalize initial word, all principal words, all words of five letters or more.

CHAPTER 13

AGREEMENT AND REFERENCE

AGREEMENT

Agreement is the relationship between a verb and its subject. A verb must agree with its subject in person and number. Problems with verbal agreement arise because collective nouns, normally singular, are sometimes regarded as plural; certain singular pronouns are frequently plural in colloquial English; and the writer fails to identify the true subject.

¶ *Each, either, another, none, neither, someone, somebody, everyone,* and other such pronouns are singular:

> None of the engineers *is* ready with his project.
> Either the electrical or mechanical engineer *is* responsible.
> Everyone may classify *his* own project.

¶ Frequently, faulty agreement occurs because the wrong word is chosen as the subject of the sentence, primarily because the subject is separated from the verb by a modifying phrase with a plural object:

> Faulty: Often a scientist such as Einstein or Newton, or others like them make a vital contribution to mankind.
>
> Correct: Often a *scientist* such as Einstein or Newton, or others like them, *makes* a vital contribution to mankind.
>
> Correct: The *equipment* — the microscopes, spectroscopes, and condensers—*was* injured by fire.

"Who" and "whom"

There is probably more confusion over the agreement of these two relative pronouns than over any other pronoun in

English. *Who* is the nominative case and is the subject of a sentence. *Whom* is the objective case and is the object of the sentence, of a verbal, or of a preposition. Whenever there is doubt of the correct form, restructure the sentence:

> Faulty: *One* of those who *is* agreeable is the president. (*Who* refers to *those.* Of those *who are* agreeable, *one is* the president.)
> Correct: One of *those who are* agreeable is the president.

> Faulty: The man *who* you wrote to is a famous mechanical engineer.

If the last sentence is examined carefully, it will be seen that *who* has no verb. The two verbs and their subjects are:

- *man is*
- *you wrote*

Who should have been *whom,* object of the preposition *to.* Here is another sentence with the same problem:

> He is one of the ablest men *who* I know.

Who has no verb—use *whom.*

REFERENCE

Reference is the term used for the relationship between a pronoun and the noun—called the antecedent—to which it is related or refers. A pronoun should agree with its antecedent in person, number, and gender:

> He ordered several books and began to read *them.*
> He had planned to interview the woman and saw *her* as soon as *she* arrived.

Some of the problems of reference are similar to those of agreement. Pronouns such as *each, anybody, anyone, someone, somebody,* and *everybody* tend to be plural when spoken, but are always singular when written:

> Faulty: *Each* of the students asked for *their* favorite breakfast.
> Correct: *Each* of the students asked for *his* favorite breakfast.
> Correct: *None* of them could find *his* coat.

Since a pronoun *refers* to an antecedent, there must be one to which the pronoun can refer. It cannot refer to a noun used as an adjective, nor to one in the possessive case:

> He was examining the man's head *who* hoped to qualify for the experiment. (*Who* cannot refer to *head*. He was examining the *head of a man who*.)
> He suffered a measles attack. *They* confined him to the house for a week. (*It confined*. . . . Measles is used as an adjective modifying *attack*—a measles attack. The *attack* confined. . . .)

Another problem with reference occurs when the pronoun refers to something implied but not stated:

> He raced for eighteen miles. *It* exhausted him.

The writer is apparently referring to the distance, which did not exhaust the runner.

> He raced for eighteen miles. *The race* exhausted him.

> He was rewarded for his contribution to the company's growth, and the company took this means of showing *it*. . . . and the company took this means of showing *its gratitude*.

This is regularly used to refer to an idea in a preceding clause or sentence:

> Several applications of the ointment were necessary before any improvement was shown. *This* confirmed his belief. . . .

There is no ambiguity in the reference of *this,* even though there is no single word to which it refers. In more formal writing, *this* would have been used to emphasize *experiment:*

> *This experiment* confirmed his belief. . . .

Problems in the reference of *this* occur when the writer uses the word to refer to concepts too far removed from the pronoun for the reader to identify the relationship simply and easily:

> Several applications of the ointment were necessary before he noticed any improvement. Now he was fully aware of the value of continued experimentation. *This* confirmed his belief. . . .

Did the *applications, awareness,* or *improvement* confirm his belief? The relationship has to be clarified.

Reference to collective and compound antecedents

If the collective noun is thought of as a unit, the pronoun is singular; if as a group of individuals or things, the pronoun is plural:

> Singular: The group of engineers elected *its* new members by voice vote.
> Plural: The group of engineers elected *their* new members by voice vote.

If a pronoun refers to a compound antecedent, the pronoun is plural:

> The boys and girls took *their* lunch.
> The boy and the girl took *their* lunch.

Either . . . or

The construction *either . . . or* normally takes a singular pronoun. However, if both antecedents are plural, the pronoun is plural. If one is plural, the pronoun agrees with the nearer antecedent:

> Either the librarian or the assistant gave *his* approval.
> Either the nuts or the bolts will be missing from *their* place in this snafu outfit.
> Either the nuts or the bolt will be missing from *its* place.

Neither . . . nor

When an antecedent is connected by *neither . . . nor,* the pronoun is singular if both nouns are singular, but plural if either is plural:

> Neither the librarian nor the assistant gave *his* approval.
> Neither the librarian nor his assistants gave *their* approval.

Note that when a pronoun refers to an antecedent with a common gender, the masculine should be used, as above.

CHAPTER 14

EDITING

THE WRITER AS EDITOR

Students in grade school were formerly, and perhaps are still, plagued with the torture called parsing the sentence. The idea was a good one, but so much emphasis was placed on how many lines to draw where and such other trivia that most students rebelled, never to parse again. The concept behind the parsing was that language, written and oral, is composed of groups of words, and that a knowledge of what these groups are and how they are related would improve comprehension of grammatical relationships and, hopefully, the ability to write. Most writers are not aware of these units of meaning—their use is second nature. But when the writer wants to improve a sentence, he must examine it, not as an inviolable unit, but as one that includes many smaller units, all of which can be altered, restructured, or removed altogether if any of these devices will improve the sentence. Writers who spill their thoughts on paper and permit the resulting splashes and blots to stand as the final form for their thoughts are guilty of dull-as-dishwater writing. All writing should be examined before it leaves the writer's desk to be certain that it is as good as possible under the circumstances. Was this clause necessary? Could a phrase have been used instead? Perhaps the modifier wasn't necessary at all! Examine the sentences and paragraphs. They should vary in length, form, and complexity. Variety is the spice of style.

The writer should be his own editor. He should

- remove deadwood
- avoid redundancy

Removing deadwood

The amount of material written every day in the United States would probably fill Pennsylvania Station, and scientists and engineers are both producing and evaluating a significant proportion of this material—usually without considering themselves as either writers or editors. Whether the technical writer has sought his destiny or had it thrust upon him, he should express himself with clarity and precision. The presence of deadwood is a serious fault in any writing, but technical writing more than any other should be functional.

Here is a report from Personnel on the basis for hiring a new applicant:

> The store of knowledge which the applicant acquired in the field of chemistry at one of the foremost institutions for research in that particular discipline was one of the factors which were used in judging his qualification for the job as a chemist with our chemical engineering department.

Imagine the irritation of the person who has to read this deadwood-filled sentence! He only wanted to know on what basis the applicant had been hired. Careful reading of this paragraph reveals only one fact—the applicant had a degree in chemical engineering from a leading university; however, even that fact is implied rather than stated. Moreover, the university is not named—surely the reader would want to know which one. Here is the sentence pared to its core:

> The applicant was hired because he had a chemical engineering degree from the University of Delaware.

Deadwood often creeps into a paper because of excessive modification. Writers often use a phrase when a word will do, a clause instead of a phrase. Different forms of modification—word, phrase, and clause—impart varying degrees of emphasis to the word modified. As a general rule, the more words used as modifiers, the more emphasis the modifier receives. Frequently the word being modified is almost obscured. The writer should be certain that emphasizing the modifier is justified, that it does not make the main concept less emphatic or focus reader attention on insignificant details. In the following series of sentences, notice the varying degrees of emphasis given the subject when it is modified by a clause, phrase, and adjective:

> The man *who has the large shoulders* hit the ball.
> The man *with the large shoulders* hit the ball.
> The *large-shouldered* man hit the ball.

The one-word modifier puts the emphasis where it belongs, on the subject and the verb. Moreover, notice the proximity of the subject to the verb.

What is wrong with the following sentence?

> There are scattered throughout the system of the stars great clouds of dust.

Several things: the logical subject is not in the subject position; there are too many modifiers in phrase form, a certain indication of deadwood; the verb *to be* is used carelessly; and the passive has crept in! Combining elements to remove deadwood and substituting an active verb eliminate unnecessary words and strengthen the sentence:

> The star system contains great dust clouds.

The passive might have been retained if the writer had wished to emphasize *dust clouds:*

> Great dust clouds are scattered throughout the star system.

The clause in the following sentence can be eliminated with no loss in sense and an increase in clarity by substituting a single adjective:

> Apparent magnitude can be measured for any star *that can be seen.*
> Apparent magnitude can be measured for any *visible* star.

And thus:

> The need for this special exercise arises from the condition *in which the animal lives.*
> The need for this special exercise arises from the animal's *environment.*

Redundancy

Redundancy is the use of excessive words to state ideas. Redundancy occurs when the writer carelessly restates a concept or when words are repeated unnecessarily. Here is a pair of sentences that should have been written as one (as shown below), eliminating the unnecessary repetition of *is:*

> Loran is an aid to navigation. The acronym is derived from the function performed—*lo*ng *r*ange *a*id to *n*avigation.

The acronym Loran derives from the function it performs—*lo*ng *r*ange *a*id to *n*avigation.

Here are several examples of pairs of sentences with the same fault:

BX cable is used to cover electrical *wires*. The *wires* carry current throughout the plant.
BX cable is used to cover the electrical wires that carry current throughout the plant.

The various components in a radio can be identified by the use of a schematic. A schematic is an outline of the parts and their function.
The various components in a radio can be identified by the use of a schematic that outlines the parts and their function.

Radio impulses can be synchronized by means of a monitoring station. Two transmitting stations have a monitoring station between them.
Radio impulses can be synchronized by means of a monitoring station between two transmitting stations.

STYLISTIC DEVICES

Parallel construction

Parallel structure, dealt with more fully earlier, is the stylistic device by which concepts similar in relationship in a sentence are structured into like grammatical units. Parallel structure eliminates the necessity for repeating subjects and verbs in many cases; in others it gives the reader a sense of the relative value of concepts by combining them with concepts of similar grammatical value. The most important point about parallel structure is that every element to be made parallel must have equal value and similar grammatical structure: if one element of the parallel structure is a prepositional phrase, all parallel elements must be prepositional phrases; if one is a clause beginning with "that," all other elements must be clauses beginning with "that," and so forth.

The following sentences waste words and obscure meaning because they were not made parallel and written as one sentence:

Darwin assumed that the variations are always small. He also assumed that many of these variations occur entirely at random. He also thought that some occur because of the purposive striving of the organism.

When these sentences are closely examined, it is evident that one verb can be used, followed by a series of parallel noun clauses used as objects:

Darwin assumed *that* the variations are always small, *that* many occur entirely at random, but *that* some occur because of the purposive striving of the organism.

Parallelism can be achieved whenever there is a series of any like grammatical elements performing similar grammatical functions in a sentence. Thus there are two elements necessary for parallel structure: similar grammatic nature and grammatic function. Subjects, objects, and modifiers in series should be made parallel:

He was determined to perform the experiment *with* economy, *with* patience, and *with* exactitude.
He was a good worker, *filing* his lab reports on time and *noting* the results of his work daily.

Faulty: *Striving* rigorously *for* a more effective style and *with* a hope *for* more precise meaning weakened his constitution.
Correct: *Striving* rigorously *for* a more effective style and *hoping for* more precise meaning weakened his constitution.

Proper subordination

Any time a sentence is revised so that concepts are stated in fewer words and with more precision, sentence elements are probably being subordinated. Properly utilized, subordination permits the writer to combine sentences and sentence elements that formerly had undue emphasis and to subordinate them to other sentence elements. Ask yourself the question, "What am I trying to tell the reader?" After the main idea has been stated, what qualifications and limitations—modifiers— are needed? Get them into the same sentence if at all possible. Don't write a separate sentence to modify a sentence element:

Faulty: The iron in the rock was dull in tone. It colored all the landscape.

Correct: The dull-toned iron in the rock colored the entire landscape.

Here is another pair of sentences in which the same fault occurs and is corrected:

Faulty: We approached the large modern laboratory. It was constructed of glass and concrete.

Correct: We approached the large modern glass-and-concrete laboratory.

In the following sentences, notice how prepositional phrases have been eliminated and the second sentence combined with the first to bring the phenomenon of the two colors into closer proximity:

Faulty: A very thin leaf of gold transmits a small amount of light of a greenish color. The light it reflects is yellow.

Correct: A very thin gold leaf transmits a small amount of greenish light, but it reflects yellow.

Avoiding excessive use of the verb "to be" (is, was, were)

Beginning writers, and many seasoned ones, frequently use the verb "to be" rather than a forceful transitive verb. Transitive verbs give thoughts emphasis, brevity, and conciseness:

The arrangement *is* good. It forces the viewer to consider the colors.

The verb *is* in the first sentence performs no function other than to link the adjective *good* to the word it modifies; however, there is no need to write a sentence simply to modify *arrangement*. Both sentences should be combined:

The good arrangement forces the viewer to consider the colors.

Here are other examples with a similar fault and correction:

The engineer *is* a major authority on building acoustics. He also pilots a private plane.

The engineer, a major authority on building acoustics, pilots a private plane.

and

> The Cabal company *is* a producer of insecticides.
> The Cabal company *produces* insecticides.

Varying sentence length and form

Don't start every sentence with a subject and verb or write only short or only long sentences. Reading monotonous sentences puts readers to sleep. Vary short sentences with comparatively long ones:

> The chemist, after filtering the paint for the eighth successive time and adding the pint of solvent to his solution, noted the result. The paint ran.

In addition to varying long and short sentences, change the sentence form. Introduce a periodic sentence (in which the predicate comes at the end rather than at the beginning of the sentence) to disrupt the constant succession of sentences beginning with a subject immediately followed by a verb. Examine your own sentences. Do they all look as if they were poured from the same mold? Since all ideas are not equal, they shouldn't look as if they had come off an assembly line.

The following paragraph varies both the form and length of sentences:

> The large coral reefs surrounding islands in the South Pacific force the islands' inhabitants to become semiaquatic. Filled with natives and loaded with food, outrigger canoes ply from one island to another. Eager fishermen, wading the shallow waters on the reef, fling nets into the fish-filled water. Straw huts nestle together against the ocean's pounding surf. The native depends on the ocean for life.

Varying the length and form of paragraphs

In a normal paragraph, the main topic, stated in the topic sentence, usually occurs in the first sentence of the paragraph. However, it is not necessary that every paragraph conform to this normal pattern. Paragraphs, like sentences, can be periodic. In a periodic paragraph, the topic sentence is the last sentence—all the details are presented, and the point to which all relate is named in the final sentence. Everyone has heard a nominating speech at a convention, during which the speaker maintains the illusion that the audience is unaware of the name of the person being nominated. The last line of the speech is really the topic sentence: "And the name of the man who promises to lower taxes, increase welfare, and work for the good of the party, is Mr. Fogbottom."

The length of paragraphs should also vary. Not every concept is of equal importance, and the paragraph's length is often a measure of the importance of the concept developed in it. However, both the form and length of a paragraph or a sentence should be functions of the nature and complexity of the material. The writer can only reinforce the tendency of the material. Emphasize the form, don't impose it.

Styling for emphasis

The following stylistic devices give main ideas force and emphasis:

- repetition
- underscoring and italics
- lists, tables, and charts
- rhetorical questions
- fragments

Repetition—The repetition of key words or concepts throughout a paragraph will force the reader to react to the main concept. Conscious repetition is a useful device for emphasis, but unconscious repetition is poor writing. Determine first if repetition will be useful; if it will, use it. Put an important term in the subject position, an important action in the verb; an idea should be structured in as many different sentence forms and lengths as possible. Use repetition, don't let it use you. Notice how often (too often) in the following paragraph on ocean currents the word *currents* is a principal part of the sentence:

> Ocean *currents* influence the weather of several countries. The Gulf *current* carries warm water from the shores of the southern United States to the shores of Ireland and England. One branch of this *current* reaches Greenland. Still another branch of the *current* affects the climate of Iceland.

Underscoring and Italics—Typographic devices should be used whenever the writer wishes to direct the reader's attention to a specific portion of the text. On the typewriter, underscore lower case and UPPER CASE, or use UPPER CASE alone. On the printed page such emphatic devices appear underscored, in **bold face,** or in *italics*. These devices should be used with caution, for overuse will only distract the reader. They will make an idea or a word jump from the page and sock the reader in the eye. Continual socking, however, will make him punch drunk.

Lists, Tables, and Charts—Tabular and graphic material is set off from the body of the paper. A series of important points in sentence form is not nearly as emphatic as a separate list printed vertically on the page. Charts and tables often convey material more quickly and easily than the printed page. No amount of verbal discussion can replace something like a road map.

Stylistic Questions—Questions the author formulates to state emphatically the main idea and to focus attention on the material are answered by the author. Since stylistic questions give the reader a handy means for finding the main idea in a paragraph or section, they should be so formulated as to aid the reader in finding that idea. If the main statement is turned into a stylistic question, the reader will have to read to answer the question:

> Why, with all the intellectual stimulation and financial advantages that a scientific career can offer, doesn't the young student choose science for a career? Because . . .

Another stylistic question, taken from a lead article in the *New York Times,* moves the reader directly into the central concern of the article:

> What is the moon made of? Within a decade American instruments and astronauts will . . .

Fragments—Occasionally, incomplete sentences, which are normally a sign of poor writing, may be used to emphasize ideas because readers unconsciously react to their difference from the normal sentence. Use sentence fragments only after you have mastered the ability to write effective *complete* sentences.

CAREFUL REVISION

The stylistic devices mentioned should be kept in mind when the paper is revised. Effective revisions are best accomplished by specific and successive readings to improve style and correct particular faults during each revision: revise to remove deadwood, to improve poor verbs, to eliminate unnecessary adverbs and adjectives. No writer is entirely satisfied with his writing, but all writers try to make their writing as good as they possibly can. All too often, technical and scientific writing seems to expound ideas with the least possible use of good style. All papers, however hurriedly written, should be revised at least once. Papers for which more time is available should be revised three times. During the first revision, examine the paper from the reader's viewpoint: have all the facts needed for accurate interpretation been included? Has the reader been carried step by step through the process? Are the level and focus proper for the reader? Studying the paper from an objective, reader-oriented viewpoint will focus attention where it belongs—on the effective transmission of ideas.

In the second revision, the writer should examine every sentence individually to determine if it contains any unnecessary verbiage. No tool is more effective here than a blue pencil. Unnecessary qualifications—adjectives and adverbs—statements beneath the reader's level, redundancies, and other unnecessary baggage should be eradicated. Get rid of phrases such as *in the field of, in the state of, seems to, began to, for the purpose of.* Examine the paper for proper subordination. Has every concept of equal importance been given equal emphasis? Have unimportant modifiers been equated grammatically with important sentence elements? Check for maximum compression! Could a concept have been expressed in a word or a phrase rather than a clause?

The third revision should concentrate on individual words, with particular emphasis on the verbs. All the main sentence elements should be examined to ascertain whether the logical subject, verb, and object are in their respective grammatical positions. Weigh each word to determine if it has the exact meaning desired; if there is any doubt, consult the dictionary or a work such as *Webster's Dictionary of Synonyms.* Is the verb active? Does it describe the action as specifically as possible?

Through all revisions, bear in mind that language is an attempt to express ideas, but it can only approximate them. The better the language, the closer the approximation.

REFERENCES AND BIBLIOGRAPHY

CITING REFERENCES IN A SCIENTIFIC WORK

Formulating general rules for footnoting and bibliographic procedures in scientific writing is difficult because the practices of leading journals differ greatly. The following rules are those most widely used among scientific writers; however, anyone writing for publication should study the practice of the journal for which the writing is intended and follow it carefully.

THE BIBLIOGRAPHY

The references in a scientific text are usually cited in relation to a numbered bibliography. And no footnotes are supplied. This bibliography functions as both footnote and bibliography.

A bibliography—sometimes cited as List of References—is a list of books with the author's name, the date and title of his work, the name of publisher, and place of publication. For an article, the entry contains: author, date, article title, journal, volume number and pages of the article. Entries in a bibliography are single-spaced, with double spaces between entries.

The author

List the author or authors, last names first, as they appear on the title page of a book or at the head of an article. For more than two authors, cite the first named author on the title page and follow with *et al.* ("and others"):

Rucklis, Hyman.
Kuds, T. E., and Korrn, F. F.
Dooper, F. O., *et al.*

Date

Immediately after the author's name give the year of the work.

Title

If the work cited is an article, capitalize only the first word. Do not underscore.

> Corwin, David R. 1962. The use of isobars in plotting cultural norms.

For a book, capitalize all words but articles, prepositions, and conjunctions. Underscore or italicize the entire title:

> Asimov, Isaac. 1961. *The Wellsprings of Life.*
> Crowther, J. G. 1931. *An Outline of the Universe.*

Publisher

If the work is a book, give the publisher and place of publication:

> Asimov, Isaac, 1961. *The Wellsprings of Life.* Mentor, New York.

For articles, underscore or italicize the name of the journal. Follow the name with a comma, the volume number in arabic, a full colon, and the inclusive pages. Where appropriate, omit prepositions and use abbreviations for the names of journals:

> Kroeber, A. L. 1939. Cultural and natural areas of native North America. *U. Cal. Pub. Amer. Arch. and Eth.,* 48: 1-242.
> Goldstein, L. J. 1947. Logic of explanation in Malinowskian anthropology. *J. Phil. Sci.,* 24: 155-166.

Arrangement

If numbers are going to be used to cite references in the text, number the entries in the bibliography consecutively. The following is an example of an alphabetically arranged bibliography:

BIBLIOGRAPHY

Alpert, Harry. 1939. Durkheim and sociologismis psychology. *Amer. J. Soc.,* 45:64-67.
Eisner, Thomas. 1962. Survival by acid defense. *Nat. Hist.,* 17: 10-18.
Gropius, Walter. 1952. *The Scope of Total Architecture.* Harper, New York.

Kardiner, Abram, and Preble, Edward. 1961. *They Studied Man*. World, Cleveland.

Weil, B. H., and Lane, J. C. 1952. Reproduction techniques for reports and information. *J. Chem. E.*, 25: 34-41.

If no footnotes had been used in the article, then the bibliography would have followed the numerical order of the references without regard for alphabetizing.

REFERENCES IN THE TEXT

References in the text are made in two ways. The author's name may be followed in parentheses by the number of the bibliographic reference.

As Alpert (1) pointed out in his study of Durkheim's social theory . . .

Muller's (5) achievement has suggested another remarkable speculation.

The date and page reference may be given in parentheses after the author's name.

As Alpert (1939, pp. 64-65) pointed out in his study . . .

Muller's (1842, p. 901) achievement has suggested . . .

It is much preferable, from a reader's viewpoint, to have the author give the page on which the material may be found. Without such information, a reader has considerable trouble locating the citation from the bibliography, which lists only inclusive pages for articles and none for books. Where practice permits, the writer should adopt that system of reference citation that best serves both his own and his readers' interest. Since the date of the study is also important in scientific and technical writing, the date is also included. When the author, date, and page reference are given in the text, no number is used either in the text *or in the bibliography*. This second system is more concise and simplifies the problem of organizing and typing the material.

COMMONLY USED ABBREVIATIONS FOR REFERENCES IN SCIENTIFIC AND TECHNICAL WRITING

Abbreviations are used for the names of well-known organizations, such as I.R.E. (Institute of Radio Engineers);

the names of fields—M.E., E.E.; and commonly used words—Inst. (Institute), Amer. (American). The following are in general use:

Bull.	Bulletin	Pub.	Publication(s)
J.	Journal	Sci.	Science
p., pp.	page, pages	Soc.	Society
pl.	plates	ser.	series
Proc.	Proceedings	vol.	volume(s)

FOOTNOTES

Because of the system of citing references in the text of a scientific work, footnotes are rarely used. The principal use of footnotes is to provide the reader with material he might like to know but which cannot be conveniently included in the body of the work. Use an arabic numeral immediately above the point in which you wish the reader to refer to the note. Skip three lines after the last line of the text and use single space for the footnotes:

> All kinds of reasoning consist in nothing but a *comparison,* and a discovery of those relations, either constant or inconstant, which two or more objects bear to each other. This comparison we may make, either when both the objects are present to the senses, or when neither of them is present, or when only one. When both the objects are present to the senses along with the relation we call *this* perception . . . . [1]

[1] From "Probability; and the Ideas of Cause and Effect," A Treatise of Human Nature, by David Hume.

Use double space between footnotes.

Do not overuse footnotes. If you think the material important enough to include, put it in the text. If it is not important enough to go in the text, perhaps it should not be included at all.

CHAPTER 16

ABBREVIATIONS

Use abbreviations sparingly. Remember the reader and his subject competency. Keep in mind these general rules before you abbreviate a word.

- Don't abbreviate short words. Spell out such words as "acre," "foot," and "mile."
- Don't use a period unless the abbreviation could be confused with another word:

      cubic yard   cu yd

but

      inch   in.

- Don't put a space between the letters of an abbreviation:

      Wrong: c u y d
        Right: cu yd

- Don't begin a sentence with an abbreviation.
- Use the same abbreviation for singular and plural.
- In case of doubt—spell out.

absolute	abs
acre	(spell out)
acre-foot	acre-ft
air horsepower	air hp
alkaline	alk
alternating-current (as adjective)	a-c
ampere	amp
ampere-hour	amp-hr
amplitude (an elliptic function)	am.
Angstrom unit	A
anhydrous	anhyd
antilogarithm	antilog

---

* This list of abbreviations is adapted from the *American Standard: Abbreviations for Scientific and Engineering Terms.*

atmosphere ....atm
atomic ....at.
atomic weight....................................at. wt
average ....avg
avoirdupois ....advp
azimuth ....az or σ
barometer ....bar.
Baume ....Be
board feet (feet board measure)........................fbm
boiler pressure................................(spell out)
boiling point ....bp
brake horsepower ....bhp
brake horsepower-hour ....bhp-hr
Brinell hardness number............................Bhn
British thermal unit................................Btu or B
calorie ....cal
candle c
candle-hour ....c-hr
candlepower ....cp
cent ....c or ¢
center to center.................................. c to c
centigrade heat unit ....chu
centigram ....cg
centiliter cl
centimeter ....cm
centimeter-gram-second (system) ....cgs
centipoise ....cp
chemical ....chem
chemically pure ....cp
circular ....cir
circular mils ....cir mils
coefficient ....coef
cologarithm ....colog
concentrate ....conc
conductivity ....cond
constant ....const
continental horsepower ....cont hp
cosecant ....csc
cosine ....cos
cosine of the amplitude (an elliptic function)................cn
cotangent ....cot.
coulomb ....(spell out)
counter electromotive force ....cemf
cubic ....cu
cubic centimeter ....cu cm or cm^3
   liquid (meaning milliliter) ....ml
cubic feet per minute ....cfm
cubic feet per second ....cfs

cubic foot ....cu ft
cubic inch ....cu in.
cubic meter ....cu m or m³
cubic micron ....cu μ or cu mu or μ³
cubic millimeter ....cu mm or mm³
cubic yard ....cu yd
current density ....(spell out)
cycles per second ....(spell out or) c
cylinder ....cyl
decibel ....db
degree ....deg or °
degree centigrade C
degree Fahrenheit ....F
degree Kelvin ....K
degree Reaumur ....R
delta amplitude (an elliptic function) ....dn
diameter ....diam
direct-current (as adjective) ....d-c
dram ....dr
dry basis ....d-b
dyne ....(spell out)
efficiency ....eff
electric ....elec
electromotive force ....emf
elevation ....el
equation ....eq
equivalent ....equiv
estimate, estimated ....est, estd
evaporate, evaporated ....evap, evapd
external ....ext
farad ....(spell out or) f
feet board measure (board feet) ....fbm
feet per minute ....fpm
feet per second ....fps
fluid ....fl
foot ....ft
foot-candle ....ft-c
foot-Lambert ....ft-L
foot-pound ....ft-lb
foot-pound-second (system) ....fps
foot second ....(see cubic feet per second)
free on board ....fob
freezing point ....fp
frequency ....(spell out)
friction horsepower ....fhp
fusion point ....fnp
gage ....(spell out or) g
gallon ....gal.

gallons per minute	gpm
gallons per second	gps
grain	(spell out)
gram	g
gram-calorie	g-cal
gram molecule	g mole.
greatest common divisor	gcd
haversine	hav
hectare	ha.
henry	h
high-pressure (adjective)	h-p
horsepower	hp
horsepower-hour	hp-hr
hour	hr
(in astronomical tables)	h
hundred	C
hundredweight (112 lb)	cwt
hyperbolic cosine	cosh
hyperbolic sine	sinh
hyperbolic tangent	tanh
inch	in.
inch(es) of mercury	in. Hg
inches per second	ips
inch-pound	in-lb
indicated horsepower	ihp
indicated horsepower-hour	ihp-hr
inside diameter	ID
intermediate-pressure (adjective)	i-p
internal	int
joule	j
kilocalorie	kcal
kilocycles per second	kc
kilogram	kg
kilogram-calorie	kg-cal
kilogram-meter	kg-m
kilograms per cubic meter	kg per cu m or kg/m^3
kilograms per second	kgps
kiloliter	kl
kilometer	km
kilometers per second	kmps
kilovolt	kv
kilovolt-ampere	kva
kilowatt	kw
kilowatthour	kwhr
lambert	L
latitude	lat or ϕ
least common multiple	lcm
linear foot	lin ft

liquid ....liq
liter l
logarithm (common) ....log.
logarithm (natural) ....log. or ln
logarithmic mean ....lm
longitude ....long. or λ
low-pressure (adjective) ....l-p
lumen l
lumen-hour ....l-hr
lumens per watt ....lpw
mass ....(spell out)
mathematics or mathematical ....math
maximum ....max
mean effective pressure ....mep
mean horizontal candlepower ....mhcp
megacycle ....(spell out)
megohm ....(spell out)
melting point ....mp
meter ....m
meter-kilogram ....m-kg
microampere ....μa or mu a
microfarad ....μf
microhenry ....μh
microinch ....μin.
micromicrofarad ....μμf
micromicron ....μμ or mu mu
micron ....μ or mu
microvolt ....μv
microwatt ....μw or mu w
mile ....(spell out)
miles per hour ....mph
miles per hour per second ....mphps
milliampere ....ma
millicurie ....mc
milliequivalent ....mequiv
milligram ....mg
millihenry ....mh
millilambert ....mL
milliliter ....ml
millimeter ....mm
millimicron ....mμ or m mu
millimole ....(spell out)
million ....(spell out)
million gallons per day ....mgd
millivolt ....mv
minimum ....min
minute ....min
minute (angular measure) ....'

minute (time, in astronomical tables) ....m
molar (of solutions)...........................M ($6M$ HCl)
mole ....(spell out)
molecular weight ....mol wt
National Electrical Code ....NEC
newton (unit of force in m.r.s. system) ....(spell out)
normal (of solutions) ....N ($3N$ KCl)
ohm ....(spell out or) Ω
ohm-centimeter ....ohm-cm
organic ....org
ounce ....oz
ounce-foot ....oz-ft
ounce-inch ....oz-in.
outside diameter ....OD
parts per million ....ppm
penny (pence) ....d
pennyweight ....dwt
pint ....pt
potential ....(spell out)
potential difference ....(spell out)
pound ....lb
pound-foot ....lb-ft
pound-inch ....lb-in.
pounds per brake horsepower-hour ....lb per bhp-hr
pounds per cubic foot ....lb per cu ft
pounds per square foot ....psf
pounds per square inch ....psi
pounds per square inch absolute ....psia
power factor ....(spell out or) pf
quart ....qt
radian ....(spell out)
radio frequency ....r-f
reactive kilovolt-ampere ....kvar
reactive volt-ampere ....var
revolutions per minute ....rpm
revolutions per second ....rps
Rockwell C-scale (hardness) ....(spell out)
root mean square ....rms
saturate, saturated ....sat., satd
secant ....sec
second ....sec
second (angular measure) ....″
second-foot ....(see cubic feet per second)
second (time, in astronomical tables) ....s
shaft horsepower ....shp
sine ....sin
sine of the amplitude (an elliptic function) ....sn
solubility ....soly

soluble ....sol
solution ....soln
specific gravity ....sp gr
specific heat ....sp ht
spherical candle power ....scp
square ....sq
square centimeter ....sq cm or cm^2
square foot ....sq ft
square inch ....sq in.
square kilometer ....sq km or km^2
square meter ....sq m or m^2
square micron ....sq μ or sq mu or μ^2
square millimeter ....sq mm or mm^2
square root of mean square ....rms
standard ....std
stere ....s
superhigh frequency ....shf
tangent ....tan.
temperature ....temp
tensile strength ....ts
thousand ....M
thousand foot-pounds ....kip-ft
thousand pound ....kip
ton ....(spell out)
ton-mile ....(spell out)
ultimate ....ult
ultra high frequency ....uhf
velocity ....vel
velocity head ....vel hd
versed sine ....vers
volt v
volt-ampere ....va
volt-coulomb ....(spell out)
watt ....w
watthour ....whr
watts per candle ....wpc
week ....(spell out)
weight ....wt
yard ....yd
year ....yr
yield point ....yp

SPELLING

No attempt has been made to formulate spelling rules in this text because such rules are dealt with at some length in most dictionaries under the heading "Orthography." When in doubt about the spelling of a word, check the dictionary and the list of commonly misspelled words that follows. If the word is on the list, chances are you have been misspelling it for a long time. Print the word plainly on a 3x5 card and refer to it frequently, trying to fix the form in your mind. If the list is used in this manner, it can perform a double function—a ready reference for frequently misspelled words and a means for learning how to spell them correctly.

absence
absorption
absurd
accept
access
accidentally
accommodate
accurate
accustom
achieve, achievement
adaptation
additionally
address
adolescence
advice (noun)
advise (verb)
aggravate
airplane
alcohol
all right
already
altar (church)

alter (change)
although
altogether
amateur
analogous
analysis
analyze
anesthetic
anxiety
apology
apparatus
apparent
appear
appreciate
approximate
arctic
argue, argument
ascend (verb)
ascent (noun)
assassin
assent (agree)
athlete, athletics

attacked
attend
attitude
attorney
attractive
audience
auxiliary
available
awkward
bachelor
balance
bargain
basis
battalion
becoming
beginning
believe
benefited, beneficial
boundary
brilliant
Britain
bureau, bureaucracy
business
busses
cafeteria
calendar
capital (city)
capitol (building)
captain
career
casualties
categories
ceiling
cemetery
certainly
challenger
champagne
changeable
characteristic, characterized
chauffeur
Christian
cigarette
collar
collegiate
colonel
color
colossal
column
comedy

coming
commit
committee
comparative
comparison
compel, compelled
competent
competition
complaint
complement (fill up)
compliment (praise)
concede
conceit
conceive
condemn
condenser
confidence
conqueror
conscience
conscientious
conscious, consciousness
consensus
consider
consistent
contemporary
contemptible
continuous
controlled
controversy
convenient
convertible
cooperative
council (group)
counsel (advice)
courteous
criticize
curious
curtain
custom
cylindrical
deceased
deceive
decent
definite
dependent
descendant
descent
describe, description
despair

desperate
develop
diaphragm
dictionary
diesel
different
difficult
dilapidated
diphtheria
disappearance
disappointment
disastrous
discipline
discretion
disease
disgusted
dissatisfied
dissipate
distributor
disturbance
divide
divine
doctor
dominant
dormitory
ecstasies
effect
efficient
eighth
eligible, eligibility
eliminate
embarrass
emphasize, emphatic,
  emphatically
encyclopedia
energetic
enforce
environment
equipment, equipped
especially
esthetic or aesthetic
exaggerate
examine, examination
exceed, excessive
excellence
except
exceptionally
excitable
exercise

exhausted
exhilarating
existence
expense
experience
experiment
explanation
extravagant
extremely
exuberance
facile
fallacy
familiar
fantasy, fantasies
fascination
favorite
February
fictitious
fiery
finally
financially
financier
foreign
forfeit
formally
formerly
forth
frantically
friend, friendliness
fulfill
fundamentally
furniture
further
government
governor
grammar, grammatically
grief
gruesome
guarantee
guardian
guidance
handkerchief
hangar
happiness
harass
height
hero, heroes, heroine
hideous
hindrance

hoard
horde
horizontal
humane
humorous
hungrily
hurriedly
hygiene
hypnosis, hypnotic, hypnotize
hypocrisy, hypocrite
hysterical
ignorance, ignorant
illiterate
illogical
imagine, imagination,
    imaginary
immediately
implement
inadequate
incessantly
incidentally
independence
incredible
indictment
indispensable
infinite
influence
ingenious
initiation
innuendo, innuendoes
inoculate
intellectual
intelligent
interest
interrupt
intolerance
irrelevant
irreligious
irresistible
irreverent
its (possessive)
it's (it is)
itself
jalopy
jealousy
judgment
kidnap, kidnaped
knowledge

laboratory
legacy
legitimate
leisurely
length
librarian
license
likable
likeness, likely, likelihood
liquor
livable
livelihood
lonely
luxury
mackerel
magazine
magnificent, magnificence
maintain, maintenance
manual
manufacturer
marriage
mathematics
mattress
meant
medicine
medieval
mediocre
Mediterranean
merely
millionaire
miniature
minute
mischief, mischievous
misspelled
mold
moral
morale
mortgage
mountainous
municipal
murmur
muscle
mustache
mysterious
naive
naturally
necessary, necessarily
Negro, Negroes
neither

nickel
niece
noticeable, noticing
notoriety
nucleus
obey, obedience
obliged
obstacle
occasion, occasionally
official
oily
omitted, omission
operate
opinion
opponent
opportunity
optimism
organization
organize
origin
outrageous
pajama
pamphlet
parallel
paralyzed
parliament
paroled
participate
particularly
pastime
pedestal
perceive
perform
perhaps
permanent
permit, permissible
perseverance
persistent
personal
personnel
perspiration
persuade, persuasion
phase
Philippines
philosophy
physical
physician
piano, pianos
pickle

picnic, picnicked
piece
planned
pleasant
pneumatic
pneumonia
politics, politician
possess
possibility
potato, potatoes
practicability
practical
practice
precede
predominant
preference, preferred
prejudice
preparation
presence
prevalent
primitive
principal (adjective and noun)
principle (noun)
privilege
probable, probably
procedure
proceed
profession
professor
program
prominent
pronounce, pronunciation
propaganda
propeller
protein
psychoanalysis, psychoanalyze
psychology
psychopathic
psychosomatic
publicly
pursue, pursuit
quantity
quantum
quarantine
quiet
quite
quixotic
really
receipt

receive
recipient
reclamation
recognition
recommend
refer, reference, referred
relative
relevant
relieve
religion, religious
remember
reminisce
rendezvous
repellent
repetition, repetitious
representative
reservoir
resistance
respectfully
respectively
restaurant
reverent
rhetoric
rhythm, rhythmical
ridiculous
sacrifice
sacrilegious
safety
salary
scandalous
scene
schedule
secretarial
seize
semester
senator
sense, sensible
sentence
separate
sergeant
severely, severity
shining
sieve
significance
similar
sincerely, sincerity
skeptical
sluggish
soccer

soliloquy
soluble
sophistication
sophomore
source
specifically
specimen, specimens
specter
speech
spicy, spiciness
sponsor
staccato
stationary (not movable)
stationery (paper)
stopping
straight
strength
strenuous
stretched
studying
substantial
subtle
succeed, success,
    successful, succession
sulfur
summary
superintendent
supersede
suppose
suppress
surprise
susceptible
suspense
syllable
symbol
symmetry, symmetrical
sympathize
synonymous
syphilis
syrup
tariff
technique
temperament, temperamental
tendency
theater
theory, theories
thorough
though
thought

thousandths
through
to (preposition)
too (adverb)
two (adjective of number)
together
traffic, trafficking
tragedy, tragic
tries, tried
truly
Tuesday
typical
tyranny
undoubtedly
unnecessary
unprecedented
until
unusual
using
usually
utensil
vacuum
varies
various

vegetables
vengeance
ventilate
vertical
vigilance, vigilantes
vilify
villain
visibility
vitamin
volume
warrant
warring
weather
weight
weird
whether
whole
woolen
woolly
writing, written
yacht
yield
zoology, zoological

CHAPTER 18

GLOSSARY OF IMPORTANT WORDS AND PHRASES

The following list is intended as a ready reference for a number of grammatical terms and some common rhetorical problems. It is by no means complete. For complete treatment of such terms, the reader is referred to Fowler, *Modern English Usage,* or Perrin, *Writer's Guide and Index to English.*

ABSOLUTE PHRASE. An adverbial modifier that lacks both the adverbial connective and the verb. Many adverbial clauses can be revised into absolute phrases that are effective stylistically:

Adverbial clause:	He experimented, *while the class watched intently.*
Absolute:	He experimented, *the class watching intently.*

Adverbial clause:	*When the job was finished,* he sighed contentedly.
Absolute:	*The job finished,* he sighed contentedly.

An absolute phrase must be capable of expansion into an adverbial clause. Check any you have written by mentally supplying them with a subordinating conjunction and a verb, as in the two adverbial clauses above.

ADJECTIVE. A word used to modify a noun. Words belonging to other classes, such as nouns and verbals, also function as adjectives.

Adjective:	The *big* man, the *important* decision, *diminutive* Delaware.
Noun as adjective:	The *stone* fence, the *cement* walk, the *glass* wall.
Present and past participle as adjective:	The *winding* stairs, the *heated* discussion.

ADVERB. A word that modifies a verb, an adjective, or another adverb. The writer can determine whether the structure he is dealing with is adverbial by asking the following questions:

- How?—adverb of manner: carefully, cautiously, greedily, loosely.
- When?—adverb of time: tomorrow, yesterday, lately, soon.
- Where?—adverb of place: downtown, uptown, downstairs, north.
- How much?—adverb of degree: plenty, enough, little, sufficient.
- Why?—adverb of cause: because, since, consequently.

Single word adverbs normally precede the word they modify, but adverbial clauses and phrases move freely. A knowledge of adverbial elements helps greatly in sentence revision.

AFFECT-EFFECT. *Affect* is a verb; it cannot be used in a noun function except in the field of psychology:

> *Faulty:* The *affect* pleased him.
> *Correct:* The *effect* pleased him.

Effect is both a noun and a verb. Use *effect* in all noun functions. Use *effect* as a verb when the meaning is "to bring about." *Affect* means "to influence."

> *Correct:* The compromise *effected* the solution.
> *Correct:* The *effect* was startling.
> *Correct:* This product will *affect* the entire market.

ANTECEDENT. An antecedent is a word or group of words to which a pronoun refers:

> The *scientist* is responsible for new products, and *he* sits on all committees.

COMPARISON OF ADJECTIVES AND ADVERBS. Adjectives and adverbs have three degrees—positive, comparative, and superlative. The positive degree is the normal form of the word: plain, round, wide, soon. The comparative indicates greater degree: plainer, rounder, wider, sooner. The superlative indicates greatest degree: plainest, roundest, widest, soonest.

The comparative and superlative of many widely used adjectives and adverbs are irregular:

good	better	best
bad	worse	worst

However, most adjectives and adverbs form their comparative and superlative degree by adding -er or -est to the positive, or by using "more" or "most":

fancy	fancier	fanciest
smooth	smoother	smoothest

or

| beautiful | more beautiful | most beautiful |
| important | more important | most important |

Problems in the use of the comparative arise when writers fail to remember that only two persons or things can be involved:

> *Faulty:* Both VTVMs can be panel-mounted, but the first is the *finest* for this use.
> *Correct:* Both VTVMs can be panel mounted, but the first is the *finer* for this use.

Many adjectives such as "round," "unique," "square," or "triangular" logically have no comparative. However, in standard English the use of comparative and superlative for such adjectives is widespread:

> He had the *roundest* face in the group.

Formal English requires that the concept be expressed as follows:

> He had the *most nearly round* face in the group.

Incomplete comparisons and superlatives are being increasingly used in advertising, but technical writers should complete all such constructions:

> *Faulty:* Dunkers is the *better-tasting* coffee.
> *Correct:* Dunkers is the *better-tasting* coffee *among our brand names*.

> *Faulty:* Smokies is the *best* cigarette.
> *Correct:* Smokies is the *best* cigarette *in the Smokies factory*.

CONJUNCTION. A word or group of words used to connect elements in a sentence. Conjunctions are both subordinating and coordinating. The major coordinating conjunctons, used to join two or more independent clauses, are

and
so
for
or
nor
yet
but

When these terms are used as coordinating conjunctions, they are always preceded by a comma.

Subordinating conjunctions introduce both adjectival and adverbial clauses. There are a great many subordinating conjunctions; some of the most frequently used are:

since
because
when
whereas
where
who
which
so
that
after
before

CONJUNCTIVE ADVERB. When certain adverbs join two or more independent clauses, they are called conjunctive. When an adverb functions in this manner, it is preceded by a semi-colon and followed by a comma. The most common conjunctive adverbs are:

moreover
accordingly
consequently
furthermore
therefore
however

The idea seemed sound; moreover, it would save both time and money.

FRAGMENT. A fragment is *part* of a sentence, mistakenly written as a complete sentence:

How long has the world existed? *Eons and eons.*

Some writers consciously use fragments as stylistic devices, generally to emphasize an idea.

GERUND. The form of the verb ending in *-ing,* used as a noun.

Swimming is fun.

A gerund, since it is a verbal, can be modified adverbially and take an object. Here is another gerund phrase used as subject:

Gerund   object   adverb
Dancing a *polka*   *well*   is difficult.

The gerund differs from the present participle, which also ends in *-ing,* only in function. The present participle is always an adjective:

Dancing the polka, the couple won first prize. (*Dancing* modifies couple.)

A gerund can also be modified by an adjective, since it has a noun function:

Good dancing is difficult.

INDIRECT OBJECT. With verbs of telling, asking, requesting, and others, an indirect object, which names the receiver of the action, is used. If the indirect object comes before the direct object, it is used alone; if it comes after the direct object, it is preceded by *to:*

> Tell *me* a story.
> Tell a story *to me.*
>
> Pass *me* the plate.
> Pass the plate *to me.*

INFINITIVE. The basic form of the verb, usually, but not always, preceded by "to": to run, to walk, to fight. With a number of constructions in English, the infinitive is used without "to":

> I shall *come.*
> I shall be happy *to come.*

Either form is correct:

> I helped him *carry* the equipment.
> I helped him *to carry* the equipment

The infinitive has both active and passive forms, and present and perfect tenses:

	Active	*Passive*
Present	To ask	to be asked
Perfect	to have asked	to have been asked

LIE-LAY. *Lie* (to recline) is intransitive:

> One *lies* down for a rest.

Lay (to place) is transitive:

> One *lays* down a book.

Principal parts:

lie	lay	lain
lay	laid	laid

LIKE-AS. In formal English, *like* is used only as a preposition, *as* as a conjunction and preposition.

> Preposition: She was as nervous *as* he was.
>   She was concerned, *like* him, over the outcome.
>
> Conjunction: He wants to win as much *as* we do.

Although *like* is being increasingly used as a conjunction to introduce comparison, writers should understand the normal distinction in usage before using *like* as a conjunction. *Like* is never used as a conjunction in formal writing.

LINKING VERBS. Some verbs perform little function other than to connect the subject and the construction following the verb. In normal usage the construction which occupies the place of an object after a linking verb is termed either a predicate nominative or adjective (according to whether the word used is a noun or adjective) or a subjective complement. Some common linking verbs are: *to be, to feel, to seem, to prove, to appear, to sound,* and *to look.* All the preceding are linking verbs when they are intransitive—that is, when they transmit no action from the subject to the object:

> The man looked *handsome.* (adjective)
> The man was a *baker.* (noun)

The writer should be careful with some verbs that are both linking and transitive verbs. Linking verbs, since they normally only state the condition or circumstance of the subject, cannot be modified adverbially:

Linking:	He *proved* wrong. (That is, he *was* wrong.)
Transitive:	He *proved* the sum. (*Sum* here is object.)
Linking:	He *felt* ill.
Transitive:	He *felt* the wall.
Linking:	He *felt* sad. (Not "sadly," because a linking verb can be complemented only by an adjective or a noun. "Sadly" is an adverb.)

Similarly, "He felt poor" (not "poorly"). We might say that a pickpocket "felt poorly"—that is, he is now in jail. He felt terrible (not terribly).

METAPHOR. An unstated comparison between two essentially dissimilar things having some quality in common. The quality belonging to one object is applied figuratively to the other:

> He *flew* to first base. (The comparison here is between the speed of the runner and the speed of a plane or bird.)
> The car *plowed* through the traffic.

The use of metaphoric verbs is common in English, and adds vividness to writing.

A metaphor differs from a simile only because a simile uses "like" or "as" (see *Simile*).

NOUN. A word that names a person, place, or thing. Nouns function as subjects and objects of verbs, objects of prepositions and verbals, nouns in apposition, and as modifiers of other nouns.

The *man* (subject) performed the *experiment* (object).

He performed the experiment with an *assistant* (object of preposition).

The father of modern communication techniques. *Alexander Graham Bell* (appositive), lived in the United States.

Modifier: *stone* fence, *brick* wall.

Noun plurals generally end in -s (but note such exceptions as "children," "oxen," "deer," and so forth). The possessive singular is formed by adding apostrophe and "s"; the possessive plural by adding "s" and apostrophe:

Singular: *boy's* books, *girl's* hats.
Plural: *boys'* books, *girls'* hats.

OBJECT. A noun, phrase, or clause that receives the verbal action; or a word or group of words functioning as the object of a prepositional or verbal phrase:

Object of sentence:	He *hit* the *ball.*
Object of preposition:	To the *house;* from the *door.*
Object of verbal:	Swimming the *lake* was his only experience.
	To run the *motor,* he needed power.

PARALLEL CONSTRUCTION. Similar grammatical units performing similar grammatical functions should be made parallel for stylistic clarity and emphasis:

The engineer assembled the equipment *by borrowing* what he could and *by buying* what he couldn't.

He was certain *that he could perform the experiment* and *that it would be a success.*

PARTICIPLE. A verb form used as a modifier. There are both present and past participles:

Present	*Past*
looking	looked
swimming	swam

The engineer, *drawing* several conclusions from the date, astonished his colleagues.

The engineer, *forced* to repeat the experiment, did so with vigor.

PHRASE. A group of words, without a subject or verb, functioning either as a noun (gerund or infinitive with their modifiers) or as a unit modifier.

Prepositional phrase:	*to* the *house; in* the *laboratory; with* the *umbrella.*
Present participle:	*Swimming* rapidly across the river, the boy caught the boat.

| Gerund—object of preposition: | By *swimming* rapidly across the river, the boy . . . |
| Infinitive as subject: | *To refine* the oil properly requires special equipment. |

PRINCIPAL PARTS OF THE VERB. The principal parts are the verb forms used to indicate tense. They are the simple—the form in the dictionary; the past—formed by adding -ed to regular verbs; and the past participle—formed by adding -ed to regular verbs.

Simple	*Past*	*Past participle*
look	looked	looked
call	called	called
shoot	shot	shot
hang	hung or hanged	hung or hanged

Note that the last two verbs do not form their pasts or past participles regularly. Whenever confusion occurs, consult the dictionary.

PRINCIPAL-PRINCIPLE. To avoid spelling errors in the use of these homonyms, remember that the *a* in principal stands for adjective; use "principal" as noun only when it means one in authority or a sum lent at interest. *Principle* is always used as a noun.

PRONOUN. A word that stands for a noun. When using a pronoun, be certain that the antecedent (the word to which the pronoun refers) is readily evident. If there is any confusion or ambiguity over the antecedent, repeat the noun.

REFERENCE AND AGREEMENT. When a pronoun is used, the word to which it refers should be clearly evident. The pronoun should agree with its antecedent in person and number. If there is any doubt about correct pronoun usage, write the sentence over, omitting the pronoun.

SHALL AND WILL. In colloquial and informal English, *will* is used with all persons to indicate the simple future; in formal English, *shall* is often used in the first person and *will* in the second and third.

Shall is frequently used to express determination:

I *shall* go.

Shall is likely to be used in the first and third person and *will* in the second when asking questions. In colloquial usage, *will* is frequently used for all persons.

When in doubt, use *will*.

SIMILE. A stated comparison, usually made with "like" or "as," between two dissimilar objects having some common characteristic.

She is light *as a feather.*
He runs *like a deer.*

SUBJECT. The performer of the action stated in the verb. The subject can be a word, phrase, or clause:

Word: The *man* hit the ball.
Phrase: *Walking four miles every day* kept him in shape.
Clause: *That he had intended to perform the experiment* is certain.

VERB. A word used in a sentence or clause, which, in conjunction with the subject and the object—if there is one—forms a complete thought. The verb indicates both the action and the time at which it occurred.

VERBALS. Verbals are parts of verbs that function as adjectives or nouns. The gerund is used as a noun, the infinitive both as a noun and as an adjective, and participles (present and past) as adjectives.

VOICE. Verbs are classified as active or passive according to whether the subject is the performer or the receiver of the action. A verb is active when the subject is the performer of the action or in the condition predicated by the verb; it is passive when the subject receives the action.

Active: The man *hit* the ball.
Passive: The ball *was hit* by the man.

Active: The car *had* already *hit* the man when we arrived.
Passive: The man *had* already *been* hit by the car when we arrived.

To form the passive, use the verb "to be" and the simple form of the verb: *was* hit, *is* hit, *has been* hit.

WHICH AND THAT. *That* refers to persons or things; it is used for all restrictive and noun clauses. It is never preceded by a preposition. *Which* is used to refer to things in nonrestrictive clauses; it is often preceded by a preposition:

The form *to which* you refer.
The form *that* you requested . . .

That is used to introduce a clause giving a reason, result, or purpose:

He shouted so *that* he might be heard.

That is often omitted both in general and formal writing when it introduces a clause:

He said [that] he would like to go.

INDEX